13 *Shale*	19 *Rock Salt*
14 *Calcareous Sand Stone*	20 *Chalk*
15 *Iron Stone*	21 *Plum Pudding Stone*
16 *Basalt*	A A *Primary Mountains*
17 *Coal*	B B *Secondary Mountains*
18 *Gypsum*	a a a *Veins*

rees Orme & Brown Paternoster Row.

Humphry Davy on Geology

Portrait of Humphry Davy by Henry Howard 1803.   National Portrait Gallery, London.

HUMPHRY DAVY

ON GEOLOGY

The 1805 Lectures for the General Audience

Edited and with an Introduction by

Robert Siegfried and Robert H. Dott, Jr.

THE UNIVERSITY OF WISCONSIN PRESS

Published 1980

The University of Wisconsin Press
114 North Murray Street
Madison, Wisconsin 53715

The University of Wisconsin Press, Ltd.
1 Gower Street
London WC1E 6HA, England

First printing

Printed in the United States of America

For LC CIP information see the colophon

ISBN 0-299-08030-7

Contents

vi   Contents

List of Illustrations

Preface

The lectures of Humphry Davy published here were transcribed from manuscripts in the archives of the Royal Institution of Great Britain, the scene of their original delivery in 1805. Like most of Davy's surviving lectures, they are written out in full and presumably were delivered as written. Since these geology lectures make up the only complete course among his surviving lectures, their publication provides for the first time the opportunity for a modern reader to sample fully the flavor of Davy's successful style of lecturing to popular, nonprofessional audiences.

At one time there existed two nearly identical sets of these ten geology lectures, one in Davy's handwriting and another in the hand of a copyist. Four of the lectures in Davy's hand are in the Archives of the Royal Institution (5, 8, 9, and 10), along with eight from the other set (all but 8 and 10). Four more in Davy's hand are in the possession of the Royal Geological Society of Cornwall in Penzance (3, 4, 6, and 7). How this latter group came to be separated from the main body of Davy manuscripts in the Royal Institution and the whereabouts of the other four lectures that must have once existed are entirely unknown to us.

Where two copies of a particular lecture still exist (3, 4, 5, 6, 7, and 9), we have closely compared their contents and established a final text by resolving the small differences in order to achieve maximum clarity of Davy's meaning. Where his intent is not clear, as in a passage crossed out in the original but retained in the copy, we have been inclusive rather than exclusive.

Because these lectures were written out in full, major editorial intervention has been unnecessary and the printed text
is essentially as Davy wrote it. The few changes have all been
made with the intent of clarifying Davy's language while preserving intact his message and his style. To further enhance the
readability of these lectures, which were given to nonprofessional
audiences, we have reduced the scholarly apparatus to a minimum,
retaining square brackets only where the supplied words determine the meaning. Without editorial indication, we have created
agreement between subject and verb where necessary and occasionally altered word order; we have modernized Davy's sometimes
archaic spelling, replacing, for example, Davy's "chrystal,"
"sienite," "Beccher," and "Leibnitz" with their modern equivalents, and we have replaced Davy's capricious punctuation altogether.

We have kept the editorial remarks to a minimum, providing
in the endnotes chiefly bibliographical information and biographical identification in addition to occasional explanatory
comments. For scientists and authors after about 1500, we have
inserted in square brackets in the text the life span and the
specific citation to the appropriate entry in the Dictionary of
Scientific Biography 15 vols. (New York, 1970-78).

Scattered through these lectures are more than one hundred
references to visual aids such as: rock and mineral samples,
drawings and paintings, and occasionally a chemical experiment.
Though in this volume we cannot reproduce more than a few appropriate drawings, we have retained Davy's clue word "Instance" so
that the reader can appreciate the emphasis and care he gave to
the visual reinforcement of his lectures.

To illustrate large-scale geological features for the lectures, Davy made preliminary sketches from which Thomas Webster
created large paintings. (See Davy's reference to this in Lecture 6, p. 90.) We had hoped that the original paintings might
still exist and that reproductions of them could be used to
illustrate these lectures, but we have gained no knowledge of
their present existence. The one survivor of Davy's preliminary
drawings in the possession of the Royal Institution is reproduced
here as Figure I.5.

We have provided thirty illustrations of geological features
related to the lectures, many of them drawings from Davy's geology notebooks. We have drawn most heavily from the notebook
Davy kept during the summer of 1804 when he traveled in Scotland
and northern England specifically to prepare for the lectures of
1805. Other illustrations have been selected from books to which
he refers in the lectures and from other appropriate works contemporary to Davy.

Madison, Wisconsin   Robert Siegfried
October 1979   R. H. Dott, Jr.

Acknowledgments

We are indebted most deeply to the Royal Institution of Great Britain for permission to publish these lectures from the manuscripts in their archives. Former members of the Royal Institution staff to whom we owe most are Jerry Weston, James Friday, and Judith Lloyd Thomas. Special thanks must be given to the present Librarian and Information Officer, Mrs. I. M. McCabe, for the promptness and patience with which she responded to our urgent pleas for belatedly requested information.

Professor Alexander Ospovat of Oklahoma State University was most generous in lending his photocopies of the four lectures in the possession of the Royal Geological Society of Cornwall so that we could compare their texts with the corresponding copies in the Royal Institution.

We thank Professor Martin J. S. Rudwick of the Vrije Universiteit, Amsterdam, for his helpful responses to several queries, especially for his suggestion that Davy's friend mentioned in Lecture 9 was George B. Greenough. And we express thanks to Mr. P. J. Gautrey of Cambridge University Library for confirming this suggestion by checking Greenough's notebooks in the Cambridge University Library.

Professor Marcia M. Goodman, History of Science Librarian at the University of Oklahoma, sent us information about eighteenth-century books not available in the Library of the University of Wisconsin.

Around home we first express our appreciation to the Graduate Research Committee of the University of Wisconsin for the financial support provided during the first summer of our work.

We owe an inestimable debt to Professor John Neu, bibliographer for the history of science at the University of Wisconsin Library, for his uncanny ability to provide clues for solving our bibliographic problems.

Barbara Richards of the University Library's Rare Book Collection cheerfully arranged for the photocopying of illustrations from valuable and fragile books.

It has been both easy and pleasurable to work with Lydia Ostenson and Elizabeth Steinberg of the University of Wisconsin Press in preparing final copy for this book, and we are grateful to them for liking Humphry Davy.

Introduction

Humphry Davy's place in the history of science is very properly secured by his important work in chemistry. He established the principle that chemical affinity is electrical in nature, and by applying that principle he was successful in decomposing a half dozen alkalies and earths and in exhibiting for the first time their constituent metals, potassium, sodium, calcium, strontium, magnesium, and barium. In 1810 he demonstrated the elementary nature of chlorine, thereby effectively destroying the validity of the oxygen theory of Lavoisier. Of lesser scientific significance but infinitely more dramatic impact, he invented the miners' safety lamp and became something of a national folk hero. The detailed accounts of these and his other chemical work have been much written about and are generally well known.

In contrast to his fame in chemistry, Davy's contributions to the history of geology are virtually unknown. Though he was serious about geology, as he was about everything that he did, he made no startling discovery in this field, and his few publications contained little of significance. But his popular lectures in geology, delivered before large and enthusiastic audiences at the Royal Institution in London, antedated by several years the appearance of the first English textbook in the field and may have had a greater influence on the development of geology than did his role in the founding of the Geological Society of London in 1807.[1]

The manuscripts from which these lectures have been transcribed are a part of the Davy collection in the archives of the Royal Institution in London. These and most of the Davy manu-

script materials came to their present location via a gift from
Sir Humphry Davy Rolleston, grandson of Humphry Davy's brother
and biographer, Dr. John Davy, to whom Humphry Davy bequeathed
his "Chemical Books and Chemical MSS."

Of the lecture manuscripts that remain, most are written
on large sheets of unlined paper, pinned or clipped together
with fasteners now long since rusted. At least one is labelled
for delivery the next day, and this appears to have been Davy's
characteristic pattern of preparation. Evidently, some of them
were used in later years, for whole pages are crossed out and
new ones inserted without the preparation of a new fair copy.
Nor, apparently, did any order decide the selection of the sur-
vivors, whose randomness prompted John Davy to say, "Their
preservation seems to have been accidental; he was careless of
them, and after they had served the purpose for which they had
been specially written, they were thrown aside, and, it may be
inferred, had not a second thought bestowed on them."[2]

The 1805 geology lectures are exceptions to the general
pattern. Not only is each lecture written in a separate bound
notebook, but they also exist in two copies, one series in
Davy's handwriting, and the other in that of a copyist. Though
each series is only partial by itself, between them all ten
lectures are available and constitute the only complete course
of Davy's lectures surviving.

The greater care given the geology lectures presumably was
intended to preserve the greater effort invested in their preparation. Geology in 1805 was a less mature science than was
chemistry, and there was not yet available, at least in English, anything like an elementary textbook from which Davy
could have drawn general information. The first text, Robert
Blakewell's An Introduction to Geology, was not to appear for
several years. In his 1839-40 edition of his brother's Collected Works, John Davy remarked that these lectures were
given at a time when "the science of Geology was in its feeble
infancy; when no one had preceded him in this country in lecturing on the subject; nor a single elementary book had been
written on it; and when he had to collect his material from
various remote sources, and from disjointed members, construct
a body of geological knowledge."[3]

The sources that Davy used, however, were not as remote
as John Davy implies. Although the Library of the Royal Institution was less than ten years old, its collection was sur-
prisingly extensive and more than adequate for Davy's needs.
Almost every identifiable source Davy used in preparing these
lectures is listed in the Catalogue of the Library of the Royal
Institution, prepared by William Harris and published in 1809.[4]
By preserving his library research in bound notebooks, Davy
could easily repeat the lectures without having to repeat the
work. He delivered geology lectures again in 1806, 1808, and
1809, presumably from the same notebooks, for there is no evidence to suggest otherwise. Though his geology course for 1811
contained long verbatim passages from the 1805 course, the

later course was much modified and reduced to only six lectures. These were given in London in the spring and in Dublin in the fall of 1811.[5]

Figure I.1. Library of the Royal Institution. From [W. H. Pyne and W. Combe], *The Microcosm of London, or London in Miniature,* with colored plates by A. C. Pugin and T. Rowlandson, 3 vols. (London, 1808-11; reprint ed., London, 1904), 3:33.

No serious analysis has been made of Davy's lectures, nor has any attempt been made to evaluate his significance as a popularizer of science. This is at least partly owing to the lack of easy availability of his lectures. Of the hundreds he gave in his eleven years at the Royal Institution, only about sixty lectures still survive in manuscript. Of these, the few published by John Davy are only a sampling and provide no co-herent corpus for systematic study.[6] The ten lectures pub-lished here constitute the only complete course of his lec-tures still in existence, and thus provide the best sample possible for an examination of his overall lecture style.

Even before Davy achieved fame for his chemical discover-ies, he had gained an extraordinary reputation for his public lectures on Albemarle Street. The *Philosophical Magazine* re-ported of Davy's very first lecture in April 1801 that the "audience were highly gratified, and testified their satis-

faction by general applause. Mr. Davy, who appears to be very
young, acquitted himself admirably well: from the sparkling
intelligence of his eye, his animated manner, and the tout en-
semble, we have no doubt of his attaining a distinguished
eminence."[7] Davy was only twenty-two when that prophecy was
made, but during the next decade, the success of his lectures,
along with his fundamental chemical discoveries, effectively
determined the pattern of the Royal Institution's activities.

During Davy's time, lectures at the Royal Institution commonly were given in sets of six to twenty, comprising courses
of general instruction in single topics. All of Davy's courses
were on scientific subjects, but others gave courses on art,
architecture, music, belles lettres, and moral philosophy.
The audiences were drawn chiefly from members of the leisure
class, men and women largely committed to the improvement of
society. They were "men of the first rank and talent--the
literary and scientific, the practical and the theoretical,
bluestockings and women of fashion; the old and the young,
all crowded--eagerly crowded--the lecture room."[8] Davy saw his
socially privileged audiences as patrons of science whose
annual subscriptions to the lectures paid the bills of the Institution and supported the laboratory where original research
could advance the understanding of nature. But it was not his
intent to make practitioners of his audience; rather he was
offering, in return for their support, the "intellectual enjoyment" that would reward their effort to understand the order
and design in nature by which everything was made subservient
to the preservation of life on earth.[9]

When these lectures were given in 1805, the science of geology was still uncertain in its foundations, and the contemporary
literature was heavily polemical. The fundamental dispute was
between the followers of Abraham Gottlob Werner of the Freiberg
School of Mines, who saw the origins of all rocks as precipitations from a primordial ocean, and the followers of James
Hutton, who claimed that a central terrestrial heat was the
chief cause of the form and nature of rocks. In the language
of the time the antagonists were known respectively as the
Neptunists and the Plutonists. With characteristic theoretical caution, Davy refused to be partisan to either side. While
he found more to praise in the Plutonic view, he saw serious
shortcomings in both systems and presented his audiences with
a critical evaluation of the current state of the science.

The accidental survival of a geological rather than a
chemical course creates benefits of genuine significance, for
Davy's credentials as a commentator on the state of geology in
1805 are impressive. From an early age on, his love of the
out-of-doors led him to collect minerals in his native, mineral-
rich Cornwall, and his later chemical studies made him a competent analyst of their contents. During his years in London
prior to 1805, he had taken at least two field trips for the
purpose of collecting minerals and making geological observations. While collecting in Cornwall in the summer of 1801, he

was joined for a time by George Greenough, who was on a similar
mission and with whom Davy later shared much of the initiative
in the founding of the Geological Society of London in 1807.
During another collecting excursion to Scotland and the Western
Isles in 1804, John Playfair accompanied Davy on his visit to
Arthur's Seat and the Salisbury Crags. Clearly Davy's geological experience by 1805 was sufficient to make his observations
on the state of the science worthy of respect, and his nonin-
volvement in the polemical disputes of the time provides us with
an account free of committed partisanship.

As noted earlier, these lectures were intended for nonprofessional audiences and contained too little detail to be equiv-
alent to a university course, but they did provide Davy's listeners with a reliable and useful understanding of the basic
information and attitudes of contemporary geology. Detailed
knowledge of geology, Davy recognized, could be essential to
the engineer, the miner, and the "improver of land," and society
as a whole could only benefit in the long run from the ever increasing fund of useful information. Davy always emphasized
what he proclaimed in the first 1805 lecture, "the general
usefulness of the knowledge of nature in increasing mental enjoyment and in strengthening and exalting our sentiments" (p.
5). There can be little doubt of his personal affirmation of
this moralistic position, nor that its appeal to his audiences
was in large part responsible for the popularity of his lectures. The spirit of a moral obligation to pursue the understanding of nature through science pervades all of Davy's public lectures, whatever their subject, and it is particularly
evident in the geology lectures of 1805. For this reason it
is appropriate that they should be preceded here by Davy's
"Introductory Lecture for the Courses of 1805," in which this
theme is explicitly and fully presented.[10]

DAVY'S LIFE AND GEOLOGICAL CAREER

Davy was born in Penzance, Cornwall, in 1778, of a family
of modest financial and cultural attainments. There were few
university graduates in the region and none in Davy's family,
though the habit of reading was a firmly established family
tradition. Davy's schooling ended when he was fifteen, after
his final year in the grammar school in Truro. Following the
death of his father a year later, in February 1795 he was
apprenticed to a physician-surgeon, John Bingham Borlase, in
his native Penzance, then a town of about 2,000 inhabitants
chiefly occupied with mining and fishing activities. During
the next three years, before he acquired his consuming interest in chemistry, Davy apparently thoroughly enjoyed his work
as assistant to Dr. Borlase; he carried out his responsibilities
with good cheer and was well liked by the patients, but there
are no reports of his having shown an unusual dedication to the
medical profession.[11]

The countryside near Penzance, only ten miles from Land's
End, exhibits much variety. Small farms and pastures give
way to barren, windswept moors, and the sea is never far
away, whether it meets the rugged granite cliffs or the gentle
beaches in coves where small streams enter. Davy knew the
Iron Age ruins on the moors and loved Mount's Bay where St.
Michael's Mount joined or left the shore with every change
of tide. His life-long love of fishing also began here when
as a child he accompanied his father or his uncle to fish the
small streams.

Figure I.2. Mount's Bay. There are several drawings of Mount's
Bay and of St. Michael's Mount in various of Davy's surviving
personal notebooks. This is the one he drew with the greatest
care. Davy MSS (15g, 121) 1805.

Davy's mind was too active not to be curious about the
geology visible all about him. His first biographer, John A.
Paris, makes the argument that Davy's interest in rocks and
minerals hardly could have been avoided in a region so visibly
full of outcrops and mining activity and relates a number of
incidents that show the young Davy to be fascinated with geology.

"How often when a boy," said Davy to me, on my showing
him a drawing of the wild rock scenery of Botallack
Mine, "have I wandered about those rocks in search of
new minerals, and, when fatigued, sat down upon the

turf, and exercised my fancy in anticipation of
scientific renown."[12]

.   .   .   .

It was his constant custom to walk in the evening to
Marazion, to drink tea with an aunt to whom he was
greatly attached. Upon such occasions, his usual companion was a hammer, with which he procured specimens
from the rocks on the beach. In short, it would appear
then, at this period, he paid much more attention to
Philosophy than to Physic; and that he thought more of
the bowels of the earth, than of the stomachs of his
patients; and that, when he should have been bleeding
the sick, he was opening veins in the granite.[13]

It is difficult to know how much of this last story to attribute
to Dr. Paris's need to achieve his witticisms, but he knew Davy
during his mature life, had practiced medicine for a time in
Penzance, and knew many of Davy's childhood friends; hence, we
are entitled to accept the substance of the story.

Davy's geological opportunities were promoted to active interests through his friendships with two men. In the winter of
1797-98, Gregory Watt, son of the famous Scottish engineer,
came to Penzance for his health. He roomed and boarded at
Davy's mother's house, Davy having rooms with Dr. Borlase.
Gregory had attended the University of Glasgow for four years
and, though he never took a degree, was clearly a very capable
young man with a knowledge and sophistication that must have
newly excited Davy about learning. They quickly became firm
friends, and their mutual interest in geology is clear from
Davy's sister's recollection that, "Mr. Watt was as one of our
family. My brother and Mr. Watt walked out together often and
they generally brought home their pockets full of stones."[14]
That their friendship was a vigorous one is obvious, as well, in
the remaining letters from Watt to Davy.[15] Only months before
his premature death in 1804, Watt published the results of some
large scale experiments on the crystallization of melted basalt
after slow cooling. Davy refers to these experiments with respect, and to the early death of his friend with deep regret,
in the seventh lecture of 1805.

The other most influential friend from Davy's youth was
Davies Giddy (later Davies Gilbert), a native of Cornwall,
graduate of Oxford, later member of Parliament, and successor
to Davy as President of the Royal Society. That Giddy's wide-
ranging interests included geology was partly the result of the
lectures given at Oxford by Thomas Beddoes on the "structure of
the earth." Giddy and Beddoes became good friends, and in 1791
they spent the summer collecting mineral specimens in Cornwall
for further illustrations of Beddoes's Oxford lectures.
Beddoes spent three years in Edinburgh, from 1784 to 1787,
just after the first appearance of Hutton's theory of the

earth, and had ample opportunity to become familiar with that
doctrine. In 1790 he published a remarkably perceptive paper
on the similarities of basalt and granite, finding general
support for the Huttonian, rather than the Wernerian, view of
the formation of the earth's rocks.[16]

Because of his radical political views, Beddoes found it
expedient to leave his position at Oxford University. He found
a more congenial atmosphere in Bristol, settling there permanently in 1793. With James Watt, he conducted a series of experiments on the production of "factitious airs," in the ultimate hope of finding some medical relief for consumption,
then so prevalent and deadly. The recently discovered physiological effects of oxygen had spawned the idea that other
artificially prepared gases might possibly have ameliorative
effect on the disease. Beddoes and Watt published the results
of their experiments in a series of pamphlets designed to gain
financial support for a Pneumatic Medical Institution where
consumptive patients could be treated. By the spring of 1798
plans had progressed sufficiently that Beddoes was looking for
a director of the laboratory, and Davies Giddy now recommended
the young Humphry Davy.

When Davy left Penzance in October 1798, he was rather
well educated, in spite of the brevity of his formal schooling.
Dr. Borlase had a modest library to which his apprentice had
been given free access, and Davies Giddy was sufficiently impressed by the precocious young man that he had allowed him
use of his own rather considerable library a few miles from
Penzance in St. Erth parish. Davy's sister later recalled
that both Davy and Giddy belonged to a subscription library
in Penzance whose books were "very select" and that, based on
what they had read, they often would argue about different
subjects. A childhood friend reported:

> He was better informed in general literature and science
> at the time he left Penzance than any person I ever met
> with on quitting the universities. He had a competent
> knowledge of history; he was acquainted with the ele-
> mentary principles of mathematics and geometry. He
> had read several metaphysical works and was read on
> the philosophy of the ancients. He read again and
> again *Brücker's History* by Enfield.[17]

Davy spent more than two years with Beddoes in Bristol,
from October 1798 to March 1801. Though his activities were
devoted mostly to his experiments with gases, where his chief
success was the discovery of the physiological effect of
nitrous oxide, or laughing gas, Davy had ample opportunity to
discuss with the versatile Beddoes many other scientific topics
besides medicine and factitious airs. We know from the records of the Bristol Public Library that Davy borrowed its
copy of Richard Kirwan's *Geological Essays* in March of 1800.
Kirwan, who had been the last serious defender of phlogiston

against the new chemistry of Lavoisier, had become one of the
most vigorous pro-Neptunist polemicists. The good Dr. Beddoes,
though a firm Plutonist, had nothing of the doctrinal spirit
in him, and it is likely that he advised Davy to read Kirwan's
work so that he might best judge the issues for himself.

Shortly after Davy's arrival in Bristol, a vein of strontianite was uncovered nearby. Since the mineral itself had only
recently been first identified anywhere, the discovery occasioned local excitement and Davy conducted some exploratory
chemical experiments on it. He discovered that the oxymuriate
of strontian (strontium chloride) has the peculiar property
of emitting light when sulfuric acid is poured into its aqueous
solution. By this time, Davy's interest in mineralogy had led
him to begin collecting specimens, and he exchanged some of
his for some from Gregory Watt's collection.[18]

Most of Davy's time, however, was occupied with his re-
searchers on nitrous oxide. Before he left Penzance, he had
discovered that the gas was breathable when mixed with large
proportions of common air, but his recognition of its exhilar-
ating effects was not made until April 1799, when he first
breathed the pure gas. From that time until the summer of
1800, he pursued further researches on the gas and its chemical relatives, work which culminated in the 1800 publication
of his *Researches, Chemical and Philosophical: Chiefly Concerning Nitrous Oxide and its Respiration.*[19] Although his
early hope that nitrous oxide might prove of lasting benefit
to sufferers of consumption and other respiratory ailments was
not fulfilled, Davy's researches established a solid basis for
his scientific reputation, and in 1801 he was offered the
position as Assistant Lecturer in Chemistry at the newly
founded Royal Institution in London.

About the time *Researches, Chemical and Philosophical* was
published, William Nicholson and Anthony Carlisle reported the
decomposition of water by means of the Voltaic apparatus, a
crude, early form of the electric battery. With characteristic enthusiasm, Davy immediately turned his attention to what
he recognized as a "new analytic tool," and by the time of his
move to London the following year, he had become the leading
researcher on chemical galvanism and had published five papers
of his work in Nicholson's *Journal of Natural Philosophy,
Science, and the Arts*. His later success with electrical
decomposition was climaxed with his preparation of potassium
and sodium, in 1807, and the metals of the alkaline earths,
in 1808.

Although brief, Davy's stay in Bristol was extremely significant in his personal development. Born and raised in rural
Cornwall, he had little opportunity for social and intellectual
intercourse until his arrival in Bristol when he was almost
twenty years old. Bristol was the focus of much literary and
social activity and gave Davy his first chance to test himself
against men and women better and more formally educated than
himself. Beddoes was an insatiably curious man, with broad

interests and a great imagination, and was well read. Davy's
first impression was that Beddoes "is one of the most original
men I ever saw,"[20] an opinion only slightly changed thirty
years later in Davy's recollection of "his wild and active
imagination which was as poetical as [Erasmus] Darwin's . . . .
He had great talents and much reading; but had lived too little
amongst superior men."[21] Through Beddoes, Davy met the novel-
ist Maria Edgeworth, who was Mrs. Beddoes's sister; Joseph
Cottle, publisher of Coleridge and Wordsworth's *Lyrical Ballads;*
William Godwin, author of utopian and radical social ideas,
and formerly husband to Mary Wollstonecraft, the early and
vigorous champion of women's rights; Peter Mark Roget, later
famous for compiling his *Thesaurus,* but then a practicing
physician in Bristol; and James Watt and Matthew Boulton, whose
Soho Engineering Works in Birmingham Davy visited to oversee
the preparation of some of the apparatus designed to prepare
and administer the gases used in the Pneumatic Medical Institution. Perhaps the most important of the acquaintances Davy
made while he was in Bristol was Samuel Taylor Coleridge, at
that time one of the truly remarkable intellects of England.
The two men became close friends, sharing, as they did, a
quickness of thought and comprehension that often left their
listeners far behind.

So Davy stood up well in this company. When Cottle asked him
how Davy would fare in comparison with the intellects of London,
Coleridge replied, "Why, Davy can eat them all! There is an
energy, an elasticity in his mind, which enables him to seize
on and analyze all questions, pushing them to their legitimate
consequences. Every subject in Davy's mind has the principle
of vitality. Living thoughts spring up like turf under his
feet."[22] And Beddoes, as well, thought Davy "the most extraordinary person I have seen, for compass, originality, and
quickness of thought," for so he wrote to Erasmus Darwin.[23]

So Davy went to London. His scientific achievements made
it fitting that he move on, his friends made it possible. It
was apparently Dr. Thomas Hope, Professor of Chemistry at the
University of Edinburgh and a friend of Beddoes, who recommended
Davy to Count Rumford as a young man suitable for Assistant
Lecturer in Chemistry at the newly founded Royal Institution.
He arrived in March 1801 and gave his first lecture on electro-
chemical topics on April 25 of that year. His success as a
lecturer was immediate, as the earlier quoted passage from the
Philosophical Magazine testifies.

As originally planned, the Royal Institution had two main
purposes: first, diffusion of the knowledge of useful mechanical inventions and facilitation of their introduction; and
second, teaching the applications of scientific discoveries
to the improvement of arts and manufacturers, through regular
courses of philosophical lectures and experiments. These aims
certainly agreed with Davy's values and were no less directed
toward a benevolent social purpose than the aims of the Pneumatic Medical Institution in Bristol had been. But Davy's oppor-

tunities for laboratory research were significantly better in
London than in Bristol, for he had been promised "the sole and
uncontrolled use of the apparatus of the Institution for private
experiments," along with any new apparatus he might need.[24] In
reality, however, his early years at the Royal Institution were
devoted more to practical science, activities more consonant
with the original purposes of the Institution, than to the pursuit of "private experiments." His work on electrochemistry,
begun so vigorously at Bristol, was now largely neglected until
1806.

In the summer of 1801 Davy was instructed to gather information for a course on leather tanning. He began his studies of agricultural chemistry in 1802. The annual lectures he
gave to the Board of Agriculture were eventually published in
1813.[25] In addition to these assigned tasks, Davy found time
in the summer of 1801 for a visit to his native Cornwall,
where he collected minerals and met George B. Greenough on a
similar mission.[26] Greenough became a member of the Royal Institution in 1804 and probably attended Davy's geology lectures of 1805.

In the summer of 1802 Davy made an extended tour of
Derbyshire and northern Wales with Samuel Purkis, a friend
gained during his studies on the chemistry of tanning. Purkis,
like so many of Davy's acquaintances, was enthralled by Davy's
knowledge and power of expression.

> We visited every place possessing any remains of
> antiquity, any curious productions of nature or art,
> and every spot distinguished by romantic and pic-
> turesque scenery. Our friend's diversified talents,
> with his knowledge of geology and natural history
> in general, rendered him a most delightful compan-
> ion in a tour of this description. Every mountain
> we beheld, and every river we crossed, afforded a
> fruitful theme for his scientific remarks. The form
> and position of the mountain with the several strata
> of which it was composed, always procured for me in-
> formation as to its character and classification; and
> every bridge we crossed, invariably occasioned a
> temporary halt, with some appropriate observations
> on the productions of the river and on the diversion
> of angling.[27]

In 1804, permission was granted to Davy to be absent "for
some weeks" from the Royal Institution for "a tour into the
North of Britain for the purpose of collecting minerals, and
of gaining information on the subjects of Geology and Agriculture."[28] It seems likely that the possibility of his de-
livering a course in geology was already planned at this date,
but there is no indication whether the idea came from the
Managers of the Royal Institution or from Davy. During his
tour of Scotland in the summer of that year, he met John

Playfair, whose *Illustrations of the Huttonian Theory of the Earth* had been published only two years earlier, and Sir James Hall, whose experiments on the controlled cooling of molten basalt constituted the first significant experimental approach to determining how rocks were formed. Apparently these two men accompanied Davy on the visit to Arthur's Seat and Salisbury Crags of which he speaks in Lecture Seven. But Davy does not mention in these lectures his visit with the same two men to Siccar Point, a few miles east of Edinburgh (see Figure I.3).

Figure I.3. Junction at Siccar Point of the Primitive and Secondary Sandstone. In the summer of 1804 Davy visited this coastal site a few kilometers to the east of Edinburgh in the company of John Playfair and Sir James Hall, the same two men who had accompanied James Hutton when he first visited it in 1788. The site has since become one of the classics in the history of Huttonian geology, for it provides dramatic evidence of the sequence of uplift, erosion, and subsidence of the earth's surface that Hutton's theory required. Although in his comments accompanying this drawing Davy noted the presence of fragments of the Primitive sandstone in the breccia which underlies the Secondary sandstone (the circles in the drawing), he made no comment on the significance of this site for the Huttonian-Wernerian debate. Davy MSS (15e, 117) 1804.

The minerals Davy collected on this trip to Scotland arrived at the Royal Institution in time to be exhibited in connection with the course of geology lectures published here. Early in January 1805 the Managers ordered that this collection

be evaluated by Mr. Hatchett who later placed their value at
more than 100 guineas. At the same January 11 meeting, the
managers also approved the preparation of a dozen paintings of
geological scenes for use in the forthcoming lectures.[29]

Davy's geology lectures began at 2 p.m. on Thursday,
February 7, 1805 and continued weekly for ten weeks.[30]   Later
in February Davy wrote to his friend Thomas Poole, "I am giving
my course of lectures on Geology to very crowded audiences. I
take a great interest in the subject; and I hope the information will be useful."[31]

In 1806 Davy found time again to take up his long-neglected
research on electrochemistry. Spectacular success was almost
immediate, for in a beautifully argued and researched paper he
offered the hypothesis that chemical attraction is electrical
in nature: "Because the strength of the galvanic apparatus
is capable of an indefinite increase, no compound body can resist its analytical powers, and we ought to be able to discover
the '*true* elements of bodies.'"[32]

During the next few years, Davy effectively put this principle into practice, and in October 1807 isolated both potassium and sodium metals by the electrolysis of their alkalies.
The dramatic qualities of these new metals, so light they floated
on water even while decomposing it with a spectacular production
of flame and explosion, only further enhanced Davy's reputation
with his audiences who perhaps understood little of the theoretical significance of their hero's discoveries. Almost immediately after conducting these experiments, Davy was taken seriously ill, and for several weeks his life was much feared for.
The Royal Institution found it necessary to place daily notices
at the door reporting his condition to the members of his loyal
audiences who came to inquire.

His recovery, though slow, was complete, and by June 1808
he reported the electrolytic preparation of the metals calcium,
magnesium, barium, and strontium. None of these had ever been
seen before and, like the alkali metals prepared the year
earlier, they reacted with water to produce heat and regenerate
the earths from which they had been prepared. These new metals
played a central role in Davy's theory of volcanic heat.

Davy's active association with the Royal Institution ended
in 1812, following his marriage to a wealthy widow. He remained
as Honorary Professor of Chemistry, without salary, but did not
pledge himself to give lectures. He wrote at that time to his
brother, ". . . I give up the *routine* of lecturing, merely
that I may have more time to pursue original inquiries, and
forward more the great objects of science. This has been for
some time my intention, and it has been hastened by my marriage."[33] He was knighted by the Prince Regent on April 8,
gave his last lecture at the Royal Institution the next day,
and married Jane Apreece on April 11, 1812. His *Elements of
Chemical Philosophy,* almost entirely based on his own work,
appeared in June of 1812. Though these few weeks represent a
complete transformation of his life, Davy did maintain an active

pursuit of "original inquiries," in the laboratory of the Royal
Institution when he was in London, in other laboratories when
it was convenient, and with his portable laboratory when he
traveled.

In October 1813 Davy embarked on a tour of the continent
that was to last nearly two years. His wife and his assistant,
Michael Faraday, accompanied him. The chief purpose of the
trip was geological, the investigation of the extinct volcanoes
in France and the active Vesuvius near Naples. The results of
his observations are discussed later in connection with his volcanic hypothesis.

Upon Davy's return to England in 1815, he was asked to solve
the problem of providing a safe lamp for use in coal mines. A
series of particularly tragic explosions had forced mine owners
to seek some kind of relief from the dangers of fire damp
(methane). By December of that year, he had devised a lamp
consisting essentially of a screen of wire gauze completely
surrounding the flame, which permitted it to be used in inflammable atmospheres without ignition outside the gauze.

Davy made another trip to the continent in 1818-19, this
time to examine the possibilities of unrolling the papyri re-
covered from the ruins of Herculaneum, buried in 79 A.D. by
the same eruption of Vesuvius that destroyed Pompeii. The
trip also gave him the opportunity again to examine the activity of Vesuvius. These and the earlier observations of 1814-15
were summarized in his 1828 paper, "On the Phenomena of Volcanoes."[34]

Beginning in 1820, Davy served as president of the Royal
Society until illness forced his resignation in 1827. His last
three years were lived under the shadow of impending death, for
in 1826 he suffered the first of a series of debilitating
strokes. He traveled for his health and wrote *Salmonia* for
amusement during his enforced inactivity, and while his health
further declined, he wrote *Consolations in Travel, or the Last
Days of a Philosopher*--as an expression of his religious views.[35]
Though written for nonscientific purposes, both books reflect
the knowledge and interests of a man whose life was spent in
scientific pursuits. The latter work, especially, contains
much scientific matter, including a brief but explicit outline
of Davy's view of the history of the earth. A few weeks after
he completed this work, he died in Geneva in May 1829.

INTRODUCTORY LECTURE FOR THE COURSES OF 1805

In addition to Davy's courses on chemistry and geology
given during the 1804-5 season at the Royal Institution, others
gave courses on natural philosophy, music, history, belles
lettres, engraving, painting, botany, and moral philosophy.
As the principal resident lecturer, Davy introduced the seasonal
offerings with a general argument for the moral value of the
scientific understanding of nature, a view that would be illus-

trated explicitly in the subsequent topical lectures. We can
only presume whether the other lecturers spoke within the same
spirit expressed in the introductory lecture, but Davy certainly did, not only in 1805 but in all of his lectures.

Davy's message in the introductory lecture is clear from
its opening sentence: "The love of knowledge and of intellectual power is a faculty belonging to the human mind in every
state of society; and it is one by which it is most justly char-
acterized--one the most worthy of being cultivated and extended."[36] The belief in a moral obligation to cultivate human
understanding expressed here is characteristic of Davy, for he
was above all else a moral man whose every conscious effort was
made in the awareness that human understanding is a gift which
we are obliged to use properly. The proper use of our in-
tellect not only produces "mental gratification" but also results in knowledge beneficial to all mankind. Some men have a
greater gift and thus a greater obligation. Davy was constantly
aware of this duty and met it with total earnestness.

The happiness of mankind, Davy argues, is directly dependent on "the progress of the inventions of the arts and the advancement of the sciences," a truth so obvious, he adds, that
it hardly requires demonstration. The romantic notion that
the untutored man of nature is the ideal of happiness is specifically denied, for not only is such a man limited by his
poverty to acts of survival, he also cannot appreciate the
"highest enjoyments" of the human mind, those "connected with
an active state of the understanding."[37]

Although convincing evidence is available to support this
contention as true about earlier times, how much more evident
it is today, Davy argues, when science and the useful arts are
so much further advanced than in any previous time. He cautions,
however, that we should not rest content with what we have
accomplished, for new knowledge of nature is potentially unlimited, and he reminds his Royal Institution audience that
to acquire such knowledge and to make it useful and available
to all, its pursuit must be patronized and thus subsidized for
the general good. Our obligations are clear: "Man is formed
for pure enjoyments; his duties are high, his destination is
lofty; and he must, then, be most accused of ignorance and folly
when he grovels in the dust, having wings which can carry him to
the skies."[38]

GEOLOGY LECTURES FOR 1805

The moral stance so explicitly expressed in the general introductory lecture persists in the geology lectures as well.
To the two themes, our obligation to understand nature and the
reward of mental gratification we will receive from the perception of providential design, Davy adds another appropriate
to the scientific content of the lectures, that of the method
for gaining knowledge of nature. Facts and facts alone are the

reliable basis for our understanding, and his corresponding distrust of theoretical speculation is the most visible characteristic of the geology lectures.

In the first analysis, the organization of the lectures is
historical. The first four trace the background of cosmologi-
cal and geological thought, myth and speculation, from antiquity
to Davy's own time. Though the writings of Whitehurst, Buffon,
Kirwan, Werner, and Hutton (all treated in the fourth lecture)
are seen by us as historical, they were in Davy's time the
sources of current knowledge and unresolved conjectures.
Davy's historical treatment of his materials shows the con-
tinuity of basic speculations on the creation of the universe
and on the structure of the earth--that there is nothing new
under the sun. From this vantage, he makes the philosophical
point that even the most recent of the larger geological writings are speculative because they still rely on an inadequate
foundation of facts. Davy's persistent condemnation of speculation is carried so far he almost suggests that to conjecture
on the origins of the earth is an impiety.

> Man attached to the globe has a limited sphere of
> action, limited faculties, and but a short period for
> the employment of them. He was not intended to waste
> his time in guesses concerning what is to take place
> in infinite duration, but he was rather born to reason
> from the past and the present concerning his immediate
> and future destinies. In philosophy, as well as in
> common life, he ought only to be guided by certainties,
> by distinct probabilities, or by strong analogies.[39]

But Davy's respect for the sacred scriptures did not include the belief that they had already provided us with the
true account of creation; on the contrary, in these lectures
he is doubly critical of those who attempt to blend their
scientific speculations with biblical literalism. He says of
Thomas Burnet's *Sacred Theory of the Earth* that it offers "one
proof amongst many others of the folly of attempts to wrest
the meaning of the sacred writings, which express general
truths, so as to make them serve as supports for hypotheses of
human invention, so as to blend them with the visions and fancies of men."[40] He criticizes Whitehurst, Kirwan, and Deluc
for similar reasons.

But Davy does not confine his criticism to those who mix
scripture and personal conjecture; speculation itself is bad,
misleading and perhaps evil in subtler ways. Specifically,
speculation subverts the proper use of our mental faculties.
Speaking of Deluc's later writings, Davy states:

> It is impossible to follow with pleasure the wanderings
> of his imagination; it is impossible not to express a
> wish that he had rather confined his active and power-
> ful genius to observations and reasonings on probabil-

ities than have suffered it to waste its strength in
vain attempts to penetrate into mysteries which have
been wisely concealed from us, and the knowledge of
which, even if it could be obtained, would be comparatively useless.[41]

Although Davy's ultimate justification for skepticism and
distrust of speculation may have been vaguely theological, its
application was a habit of mind so strong and conscious that it
dominated the structure and content of the lectures. His evaluation of every geological writer was determined by how closely
the writer adhered to facts and confined his speculations to
analogies from them. Davy's comments on Plato in Lecture Two
represent his clearest criticism of this sort.

> Plato appears to have attended to the observations
> of nature less than any of his contemporaries. . . . He
> has thrown new clouds of abstracted metaphysics over
> doctrines originally obscure, and in examining the
> work, the mind is continually perplexed with fancies of
> numbers passing into quantities, ideas becoming matter,
> and the dreams of the [philosopher] forming the universe
> after the model of his own intellectual world and
> fashioning the mind of man after the image of the
> material world.
>   The eloquence of Plato, even in his abstracted
> systems of the universe, is of a high character. It may
> delude and seduce, but when the beauty of language is
> taken away, the charm vanishes. He promises a reality,
> he presents a dream. His theories, like brilliant
> clouds of the evening, may delight and affect the
> imagination, but they present nothing permanent. They
> cannot be touched, they cannot be retained, and they
> vanish without leaving in the mind a trace of their
> existence.[42]

Most of the ancient writers are criticized for their seem-
ing belief that rational thought alone is adequate to develop
paths in practical as well as abstracted science. However,
Davy recognizes that some of the ancients did appeal to experience and observation, "and it is to . . . this alone that we
owe whatever is found true or valuable in their reasonings on
sensible objects, on nature, and on the earth."[43]

To Davy, then, progress in science clearly is made in
proportion to an increasing reliance on detailed observation
and experimentation. It was Francis Bacon who had established
"the sciences upon their true and immutable foundations," for
he was the first to assert that "all the sciences could be
nothing more than expressions of facts and that the first step
towards the attainment of real discovery was the humiliating
confession of ignorance."[44] Robert Hooke was one of the first
geological thinkers to inquire of "the theory of the earth with

enlightened views and distinct plans," and although he did
speculate to some degree, Davy thought it important to point
out that Hooke "seems, however, to have restrained his imagination. He contented himself with demonstrating a certain part
of the existing order of nature without attempting a general
system of the past changes of the globe and its future destiny."[45]

Chemistry and mineralogy seems to Davy to provide the factual basis necessary to the progress of geological understanding.
He credits Whitehurst as among the first to utilize chemical
knowledge in his attempts to understand the present appearance
of the earth. But Whitehurst's speculative writings seemed to
Davy only to provide further evidence of the insufficiency of
human reason to account for the "primeval state of things."[46]

Buffon's account of the origin of the earth as resulting
from a collision of the sun with a comet appeared to Davy to be
among his wilder speculations, lacking any direct supporting
evidence whatsoever. Even Buffon's presumption of the molten
state of the earth was also pure speculation, Davy argues, for
the greater number of rocks are

> wholly incapable of being fused by any heat that we
> are able to apply to them. And though it is not im-
> probable in a temperature such as that of the sun they
> would become liquid, yet we have no right to make in-
> ference merely for the sake of forming a theory. For
> a theory ought in all cases to be not an expression of
> conjecture but an arrangement of facts.[47]

Nor did the experiments of Sir James Hall on the melting and
slow cooling of basalt persuade Davy to allow the molten condition as possible for all rocks, for "no strong analogies can
be applied, from these most interesting facts, to granite, to
the regularly formed gems, or to any of the perfectly aggregated rocks containing no organic remains."[48]

Although Davy first mentions the Neptunian view of Werner
and the Plutonic view of Hutton in Lecture Four, he gives a
fuller evaluation of them in Lecture Five. He holds that both
systems are founded principally on conjecture and finds neither
wholly convincing, "for as I have already said, their principal
materials are neither fusible by fire nor soluble in water."[49]

Davy cites the example of Newton as one who "with true
sagacity" avoided conjecture and contented himself "with the
arrangements of known phenomena and the explanation of nature
by analogy compared with facts. And in consequence he had the
glory of having discovered some of the most extensive and most
sublime of physical truths."[50] If Dr. Hutton and Mr. Werner,
Davy adds, had done likewise, they would have deserved a much
higher praise than merely to be remembered as the founders of
ingenious hypotheses.

> But when attempting to explain appearances, they attri-
> bute to agents powers which they have never been ob-
> served to exert, or refer effects to causes, the opera-
> tion of which they are ignorant, their suppositions do
> not merit the name of science.[51]

In spite of this "pox on both your houses" attitude, Davy
had a slight preference for the Huttonian conjectures over the
Wernerian errors, saying that, "if the Huttonian or Plutonic
theory of the formation of primitive rocks be founded upon conjectures, the Neptunian geognosy of Werner and Kirwan is almost
wholly built upon error and chemical impossibilities."[52] However
slight his Huttonian preference appears here, it was real, and
Davy never took the Neptunian theory seriously. In later years
his preference for the igneous origin of primitive rocks became
a clear commitment even while he rejected the Huttonian steady-
state view of the earth's history.

In the fifth lecture Davy makes clear the distinction between primitive and secondary rocks, excluding from the primitive classification any rock containing forms of animate life,
strata deposited from volcanoes, and alluvial deposits of recent origin. Most of this lecture is concerned with the description of primitive rocks, both at the detailed compositional
level, which he illustrated by hand samples, and at the larger
aspect level, illustrated by means of sketches prepared for
the purpose. Toward the end of the lecture, Davy notes that
primitive rocks typically make up the core of mountain chains
and takes the opportunity to note the role that mountains play
in the happy design of the earth that promotes the welfare of
man. Though the detailed knowledge exhibited is naive and reflects a general ignorance of the geography of North America,
he sees that the mountains determine the climates of their
neighborhood by providing cool air from their summits and by
blocking harsh winds with their masses. Still more important,
he recognizes, is the water mountains supply, their coldness
condensing the moisture from the air. The water and ice thus
produced are the sources of springs and streams that fertilize
the valleys below.

The sixth lecture opens with a brief account of the early
views on the nature of fossils, whose presence chiefly charac-
terizes the secondary rocks. Davy credits the admirable Robert
Hooke as one of the first "who fully demonstrated" that fossils
"were truly organic remains and the relics of living beings. . . ."[53] He relates, as well, other characteristics that
distinguish secondary from primitive rocks. After describing
limestone, sandstone, soft schists (shale), and pit coal, he
turns to the "last great class of the secondary masses . . .
whin, or basalt."[54] Today basalt is recognized as entirely igneous in its origin and does not belong with Davy's secondary
rocks which otherwise we classify as sedimentary. Davy's error

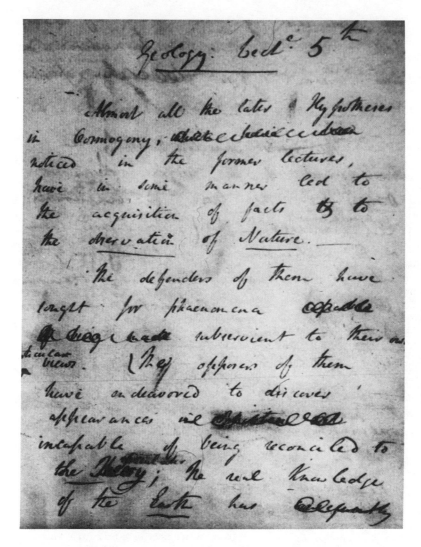

Figure I.4.   First page of the fifth lecture of 1805 in Davy's handwriting.

Figure I.5. Aspect of Primitive Rocks. Sketch for a Large
Painting for next Thursday. This is the only surviving sketch
of those Davy made for the 1805 lectures. It was preserved by
Michael Faraday who bound it in the Royal Institution copy of
John Ayrton Paris's *The Life of Sir Humphry Davy, Bart.* (London,
1831), at page 130. Although this is an idealized sketch, Davy
included the real Ben Nevis in the distant mountain scenery of
the background. Other writings on the sketch are Davy's instructions to the artist, Thomas Webster, and read: "Granite,
grey, brown; Quartz, stratified, white; Serpentine, small imperfect blocks, green and red; Porphyry, distant blocks, small,
red; Schist, irregular strata, grey; Marble, stratified indis-
tinctly, grey, blue; Sienite, irregular but perfect blocks,
grey, yellow."

here was not the result of his personal confusion only, for the
proper classification of basalt had long been a point of major
dispute. By the end of the eighteenth century, many basaltic formations had been clearly associated with the lava flows of extinct volcanoes. However, others had been found in layers
between rocks which were clearly of aqueous origin, or as dikes
cutting across other layered formations. But Davy deferred
his discussion of the origin of the secondary rocks to Lecture
Seven.

In describing in Lecture Six the appearance of basalt,
Davy indicates that it is made chiefly of hornblende and feldspar, but "not so perfectly crystallized as in syenite," a
species he previously treated under primitive rocks.[55] He emphasizes the variable appearance of basalt and spends some time
describing its columnar forms at Fingal's Cave on the island of
Staffa, which he had visited the previous summer, and at the
Giant's Causeway, which he did not see for himself until the
summer of 1806. The interlayering of basalt with other
secondary strata is also noted as appearing in the mines of
Durham, Northumberland and Cumberland.

In the opening remarks of Lecture Seven, Davy accurately
paraphrases Huttonian actualism by stating that an understanding
of geological events of the past "must be from the examination
of existing operations that produce appearances similar to those
which are the objects of enquiry."[56] He recognizes the aqueous
origin of the secondary rocks even while noting the evidence
adduced to support an igneous origin. Davy is uncertain only
about basalt whose peculiar properties and occurrences seemed
to defy explanation by either aqueous or igneous origin. He
states that specimens closely resembling volcanic lavas have
been cited as evidence of its igneous origin, but reports, as
well, that basalt often occurred in layers between rocks of
aqueous origin, such as sandstone and limestone, which showed
no evidence of having been subjected to temperatures needed to
melt the basalt. Davy concludes that the origin of basalt remained undecided at this time.

Davy's treatment of metallic veins in Lecture Eight is
descriptive and based chiefly on his personal familiarity with
the veins found so extensively in his native Cornwall. He
points out that Werner and Hutton had been no more successful
in explaining the origin of metallic veins than were earlier
writers and emphasizes that an explanation will come only with
an advance in the chemical knowledge of mineralogy.

Volcanism is the last major topic in these lectures. Since
in 1805 Davy had never seen a volcano, he fills a major portion
of Lecture Nine with readings from the writings of early observers of Mt. Etna and Mt. Vesuvius. Following the readings,
he begins a discussion of some of the hypotheses previously
offered to explain the origin of volcanic heat. He first rejects the Huttonian idea of a central heat, for if such a condition had existed in the earth for a long period of time,

there would not be "not a few widely scattered volcanoes but
one ignited and glowing mass."[57]
    Although Davy did not attempt a "best" explanation of vol-
canic heat, he clearly favored the combination of pyrite and pit
coal, a hypothesis he develops at some length in the first half
of Lecture Ten. In a proper state of subdivision, he explains,
the pyrite will take fire spontaneously and ignite the pit coal,
which will then provide the great heat necessary to melt the
adjacent rocks and thus form the lava. He then adds with
characteristic caution:

> With regard to the various hypothetical opinions
> that have been advanced, I hope it will not be con-
> ceived that I have placed any indiscreet confidence in
> them. In entering upon speculation, the principal
> object in my view was to develop the general facts
> known with regard to volcanic eruptions, in a connected
> and analogical order. . . .[58]

Before any "perfect explanation" can be achieved, much more
factual information is needed. And though the difficulties of
determining the behavior even of familiar materials under con-
ditions that occur inside the earth may seem insurmountable,
yet this should not be cause for discouragement, for equally
difficult problems of the past "have been enlightened by the
discoveries of modern science," and the

> investigation of natural causes is always a happy ex-
> ercise for the human understanding, not a gratifica-
> tion of idle curiosity, but of the love of useful know-
> ledge. And the development of truths of this kind is
> of the highest interest, displaying at the same time
> the talents of man, the majesty and variety of nature,
> the wisdom and perfection of the laws of nature.[59]

    In the final portion of this last lecture, Davy returns to
the theme of providential design. Volcanoes, though appearing
as accidents in the general scheme of nature, yet "bear a distinct subservience to the general harmonious series of natural
operations."[60] The lecture finishes with an extended series of
illustrations establishing this point and he concludes,

> Amidst the various infinitely diversified changes of
> things, nothing can be said to be accidental or with-
> out design. Even the most terrible of the ministra-
> tions of nature in their ultimate operation are preg-
> nant with blessings and with benefits. Beauty and
> harmony are made to result from apparent confusion,
> and all the laws of the material world are ulti-
> mately made subservient to the preservation of life
> and the promotion of happiness.[61]

DAVY'S LATER GEOLOGICAL ACTIVITIES

We know little of Davy's other geology courses before 1811,
and it is reasonable to presume that they were essentially identical with the 1805 course. John Paris has left a brief account
of the 1808 geology course, following his account of Davy's
electro-chemical lectures of the same year.

> His evening lectures on Geology were equally attrac-
> tive; and by a method as novel as it was beautiful,
> he exhibited, by the aid of transparencies, the struc-
> ture of mountains, the stratification of rocks, and the
> arrangements of mineral veins.[62]

As we shall see below, in 1808 Davy also introduced a new
hypothesis for the origin of volcanic fire.

We know a great deal more about the geology lectures of
1811. Five of the six given still survive in the archives of
the Royal Institution, and the *Philosophical Magazine* published
an extensive abstract of their contents shortly after Davy delivered them.[63] Thomas Allan, Davy's friend and the publisher
of the *Caledonian Mercury,* published an even more extensive
account of the lectures in a fifty-three page pamphlet.[64] The
reduction from ten to six lectures was accomplished largely
by the omission of most of the historical material from the
first three lectures of 1805 and by a general consolidation of
the material in the remaining ones. Davy's providential theme,
that all nature conspires to preserve "the beautiful cycle of
terrestrial events . . . subservient to the permanency of
life,"[65] is more clearly and forcefully visible than in the
1805 lectures. In particular, he presents the erosion and degradation of rocks and mountains as serving chiefly to maintain the production of new soils and new fertility necessary to
the welfare of mankind. This emphasis, of course, seems right
out of the writings of James Hutton, but it also must have been
important to Davy because of his continuing agricultural experimentation connected with his lectures to the Board of Agriculture.

His attitude toward the Plutonist-Neptunist debates had
not changed; the lectures of 1811 show a still cautious but
distinct preference for the agency of fire in the formation
of primitive rocks, but without abandoning the significance of
water in the degradation of hills and mountains. The following
passage from his letter to Thomas Allan in response to Allan's
abstract of the 1811 lectures gives the clearest statement of
Davy's position.

> In my next course of geology I shall modify many
> of the doctrines, and certainly mould the whole some-
> what more into a plutonic form. I think you make me
> more unjust than I conceive I was (certainly more un-
> just than I intended to be) to Dr. Hutton, and his en-

lightened and powerful philosophical defenders. My
leaning has always been to fire, even before I discovered the metals of the earths; but I think the
agency of water must be likewise introduced to explain
the siliceous formations (water highly heated under
pressure).

In my course through Herefordshire and Shropshire
a few days ago, I witnessed some good plutonic facts--
hummocks, and hills of trap and win [whin], rising in
the midst of limestone strata, almost vertical near
them, and gradually recovering their parallelism towards the remote parts. The Huttonian doctrine, considered as an hypothesis, has many advantages over all
the other views; it offers probable explanations of
many more phenomena, and presents fewer difficulties;
difficulties, however, it still has, and they must be
removed before it can be considered as a genuine
theory.[66]

Davy's chief contribution to the Neptunist-Plutonist debates, though it came too late (1822) to have much persuasive
significance, was his paper "On the State of Water and Aeriform
Matter in Cavities Found in Certain Crystals."[67] The specimens
the title refers to were either "rock crystals or other siliceous stones." Davy undertook an analysis of the contents of
these cavities because it occurred to him that "these curious
phenomena might be examined in a manner to afford some important
arguments as to the causes of the formation of the crystal."[68]
He may have had it in mind to refute the kind of argument
offered by the Neptunist, Edward Daniel Clarke, Professor of
Mineralogy at Cambridge, about whom John Paris reported:

I well remember with what triumph the late Dr.
Clarke in his popular lectures in Mineralogy at
Cambridge, paraded a fine crystal containing water in
its cavity. "Gentlemen," said he, "there is water
enough in the very crystals in my cabinet to extin-
guish all the fires of the Plutonists."[59]

Placing the crystals under a liquid (either water or mercury), Davy drilled into these small cavities with a diamond
drill. He invariably found that the liquid entered the cavity
thus indicating a gaseous pressure below that of the atmosphere.

It appears difficult to account for the phenomenon,
except on the supposition of their being formed at a
higher temperature than that now belonging to the surface of the globe; and the most probable supposition
seems to be, that the water and the silica were in
chemical union, and separated from each other by
cooling.
. . . I shall conclude by observing, that a fact which

has been considered by the Neptunists, above all other,
as hostile to the idea of the igneous origin of crystal-
line rocks, namely, the existence of water in them,
seems to afford a decisive argument in favour of the
opinion it has been brought forward to oppose.[70]

But the geological problem that most persistently challenged
Davy's interest and attention was the source of volcanic heat,
and this is evident in his revised geology lectures of 1811.
Here he rejects the idea of a permanent central heat in the earth
for the same reason that he had rejected it earlier and repeats,
nearly word for word, his statement in the ninth lecture of 1805.

. . . if the interior of the globe had been from all
time in a state of ignition, the effects must have been
long ago communicated to the surface, which would have
exhibited not a few widely scattered volcanoes, but one
glowing and burning mass.[71]

It is perhaps not surprising that Davy, as a chemist, would
be partial to a chemical explanation for volcanic heat. But
the preferred fuel of the 1805 lectures, pyrite and pit coal,
is replaced in 1811 by the active metals Davy first prepared
from the alkalies and the alkaline earths, materials long known
as major constituents of the common minerals. As explained
before, the remarkable properties of the new metals made their
discovery one of Davy's most spectacular achievements. But
even in 1807 when Davy first reported the production of potassium and sodium, he hinted at the implications their properties
would have for geology, seeing them "as likely to lead to
numerous discoveries relative to the formation of various
stones, strata, and mountains."[72] The hints of 1807 became
somewhat more specific the next year when he reported the
preparation of the four metals--calcium, magnesium, strontium,
and barium--from the alkaline earths.

The metals of the earths cannot exist at the surface of
the globe; but it is very possible that they may form a
part of the interior, and such an assumption would offer
a theory for the phenomena of volcanoes, the formation
of lavas, and the excitement and effects of subterranean
heat,* and would probably lead to a general hypothesis
in geology.

Davy further explains the implications of his assumption in the
note he adds to this passage.

*Let it be assumed that the metals of the earths and
alkalies, in alloy with common metals, exist in large
quantities beneath the surface, then their accidental exposure to the action of air and water, must

produce the effect of subterranean fire, and a produce
of earthy and stony matter analogous to lavas.[73]

In the first geology lecture of 1811, this assumption is
developed more fully and offered as a solution to the Huttonian
problem of explaining the source of the central heat the
Plutonic theory required.

> This may be accounted for by supposing the interior of
> the globe composed of the metals of the earths, which
> the agency of air and water might cause to burn into
> rocks; and even the re-production of these metals may
> be conceived to depend upon electrical polarities in
> the earth; and in this manner an harmonious order may
> be assumed: but though the idea is one which I have
> myself ventured to throw out, I cannot avoid saying
> that it rests on pure speculation. It does not command
> our assent, nor has it for me that high degree of pro-
> bability which necessarily induces conviction.[74]

He returns to this hypothesis again in the sixth lecture of
1811, which deals with volcanic action.

> Electricity has been the great agent for producing
> metals from earths and alkalies, and if it can be
> supposed that electricity acting under the pressure
> of the ocean or in the great subterraneous laboratory
> of the mineral kingdom is capable of separating the
> inflammable bases of the earths from their oxygen,
> then there might be assumed a perfect equilibrium, a
> consistent balance of powers in the system of ter-
> restrial change . . . .[75]

Davy then offers other examples of natural equilibria which, by
analogy to his conjectural one, offer support for its probability. Both the oxygen-carbon dioxide cycle between plants and
animals, and the hydrologic cycle of ocean to atmosphere, to
rain, to streams, and back to the ocean, seem to allow the possibility that his own cycle is at least reasonable to hypothe-
size.

   Davy's lecture demonstrations of his active metal hypothesis as the source of volcanic heat must have been spectacular.
John Paris reports:

> I remember with delight the beautiful illustration of
> his theory exhibited in an artificial volcano constructed
> in the theatre of the Royal Institution. A mountain
> had been modelled in clay, and a quantity of the metal-
> lic bases introduced into its interior; on water being
> poured upon it, the metals were soon thrown into violent
> action--successive explosions followed--red hot lava

was seen flowing down its sides, from a crater in min-
iature--mimic lightenings played around, and in the
instant of dramatic illusion, the tumultuous applause
and continued cheering of the audience might almost
have been regarded as the shouts of alarmed fugitives
of Herculaneum or Pompeii.[76]

Davy was not one to let his speculations lie untested if
there was any way to gather the relevant evidence. In 1813 he
undertook an extended visit to the continent with the intent
of gathering, in France and in Italy, this kind of evidence
for his volcanic hypothesis. Such a trip during a time of war
between England and France required special consideration.
John Paris describes the venture:

After the Emperor of the French had sternly refused his passport to several of the most illustrious
noblemen of England, it was scarcely to be expected
that Sir H. Davy would have been allowed to travel
through France, in order to visit the extinct volcanoes
in Auvergue and afterwards to examine that which was
in a state of activity at Naples.
No sooner, however, had the discovery of the decomposition of the alkalies and earths, and its probable
bearings upon the philosophy of volcanic action, been
represented by the Imperial Institute to Napoleon,
than . . . he immediately and unconditionally extended the required indulgence.[77]

Davy and his party left England in October and journeyed
directly to Paris where he received a most cordial welcome from
the French scientific community. When he left Paris at the
end of 1813, the cold winter may have persuaded him to pass up
any inspection of the extinct volcanoes of Auvergne, for he
traveled rapidly to Montpellier where he had the time and
better weather to examine the geology of southern France. In
a letter to his brother John written in March 1814 from Florence,
he briefly summarized his observations.

We have made a most interesting voyage in event-
ful times. I have passed from the Pyrenees to the Alps,
and have twice crossed the Appenines, and have visited
all the most remarkable volcanoes in the south of France.
All the basalt that I have seen between the Alps and
Pyrenees is decidedly of igneous origin. I have observed some facts on this subject that are, I believe,
new--a regular transition of lava into basalt, depending upon the different periods of refrigeration, and
true prismatic basalt in the interior of an ancient
lava.[78]

It seems likely to suppose that this experience would have
settled the basalt problem for Davy, and indeed he implied so
in his first presidential address to the Royal Society in 1820
where he summarized the current state of the science of geology.

> As to the origin of the primary arrangement of the
> crystalline matter of the globe, various hypotheses
> have been applied, and the question is still agitated,
> and is perhaps above the present state of our knowledge;
> but there are two principal facts which present analo-
> gies on the subject; one, that the form of the earth
> is that which would result, supposing it to have been
> originally fluid; and the other, that in lavas, masses
> decidedly of igneous origin, crystalline substances,
> similar to those belonging to the primary rocks, are
> found in abundance.[79]

Yet the problem was not resolved, for not all basalt could
be identified with lava flows, and its occurrence between layers
of secondary rocks still seemed to Davy to warrant his 1811
classification of it as secondary and non-igneous. The proper
identification of this troublesome rock forever eluded Davy,
and in 1823 he wrote his cousin in Dublin, Edmund Davy, of his
continuing puzzlement, "I have been visiting some of the wild-
est spots in Mayo and Donegal, and have again and again been
studying the mysterious basaltic arrangements of Antrim; but I
almost despair of any adequate theory to account for the
phenomena."[80]

This is the last word we hear from Davy regarding the basalt problem, which was itself incidental to his interest in
volcanoes. His theory that volcanic heat might derive from the
action of subterranean waters on the metals of the earths and
alkalis had taken him to the continent in 1813, and before his
return to England in May of 1815, he visited Naples and studied
Vesuvius on two occasions. No reports of these observations
appeared until 1828 when he was again traveling on the continent, this time in a final, futile effort to regain his health.
The paper was based on both his 1814-15 observations on Vesuvius
and his much more extensive ones made in 1819-20.[81] The chief
purpose of his experimentation had been to determine the chemical nature of the immediate products of volcanic activity, especially the nature of the gases emitted, and whether any chemical combustion was taking place. His evidence was strongly
negative regarding chemical combustion, for nitre thrown upon
the hot lava stream produced no visible change in its appearance,
nor was any oxygen consumed by a bit of molten lava placed in a
closed bottle and allowed to cool.

Davy's conclusion to this paper is characteristically
cautious and reflects an almost wistful attachment to his own
theory of metallic combustion in spite of the lack of support-

ing evidence.   "It appears almost demonstrable that none of the
chemical causes anciently assigned for volcanic fires, can be
true."[82] He specifically mentions the combustion of pit coal,
the cause he had preferred in his lectures in 1805, and Lemery's
sulphur-iron mixture described in Lecture Nine.   But he takes
care to show that the evidence does not preclude his theory of
the oxidation of the metals of the earth and alkalies, and cites
some circumstances that seem to favor it: specifically, the established existence of subterranean cavities in the neighborhood of volcanoes (near Vesuvius at least), the general proximity
of the sea to major volcanoes, and the production of common salt
from active volcanoes.

> On the hypothesis of a chemical cause for vol-
> canic fires, and reasoning from known facts, there
> appears to me no other adequate source than the oxi-
> dation of the metals which form the bases of the earths
> and alkalies; but it must not be denied that consid-
> erations derived from thermometrical experiments on
> the temperature of mines and on sources of hot water
> render it probable that the interior of the globe
> possesses a very high temperature; and the hypothesis
> of the nucleus of the globe being composed of fluid
> matter, offer a still more simple solution of the
> phenomena of volcanic fires than that which has just
> been developed.
> Whatever opinion may be ultimately adopted on
> this subject, I hope that these inquiries on the actual
> products of a volcano in eruption will not be without
> interest for the Royal Society.[83]

On this inclusive note Davy ended his active geological
career.   The problems which had interested him most, the source
of volcanic heat and the nature of basalt, had not been solved
to his satisfaction.   During the last two years of his life,
he was too ill to undertake geological excursions, but he
traveled much in the mountains of Switzerland, Austria, and
Italy during the summer months and moved to the warmer clime
of Rome during the winter.   The thoughts inspired by this
spectacular scenery and his reflections on a life spent in
science are blended in his last and posthumously published work,
Consolations in Travel, or The Last Days of a Philosopher.   The
work is essentially a theologically based justification of a
life spent in science, and Davy believed that it contained
truths which would "be extremely useful to the moral and philosophical world."[84] As an essential part of the book's purpose,
he included an account of the geological history of the earth.

Davy did not force his geology to conform to scripture;
he believed that the revealed truth of the sacred writings and
the truth to be found in nature through scientific investigation formed a general harmony, rather than a detailed corres-
pondence.   Yet he had no doubt that science properly founded on

its own observations and guided by its own methods ultimately
would prove to be consistent with the general truths of revel-
ation. He obviously sensed a confirmation of that faith in
the facts reported in a major paper by William Buckland in
1822.[85] Buckland's paper described the bones of more than
twenty, mostly extinct species of animals found in a jumbled
array in a cave in Yorkshire, which Buckland thought gave testi-
mony to the geological truth of the universal deluge recounted
in the Bible. The Royal Society awarded Buckland its Copley
medal for the paper and Davy, as President, made the award
speech. "It is gratifying to feel that the progress of
science establishes, beyond all doubt, the great catastrophe
described in the sacred history, and the account of which is
blended with the tradition of so many ancient nations . . . ."[86]
Even in his enthusiasm Davy's position was that science had independently confirmed scriptural truth, not that science should
conform to it. In the *Consolations* he expresses this position
again: "If the scriptures are to be literally interpreted and
systems of science found in them, Galileo Galilei merited his
persecution, and we ought still to believe that the sun turns
round the earth,"[87] and a little later, "I have made use of the
term diluvian, because it has been adopted by geologists, but
without meaning to identify the cause of the formations with
the deluge described in the sacred writings . . . ."[88]

In spite of the care which he took to distinguish between
science derived from scripture and science which confirmed its
truths, it is clear that Davy wanted to believe that science
and scripture would ultimately prove to be compatible. That
desire may even have affected his attitude toward some aspects
of the Huttonian theory, though the scientific basis for his
reservations was solid enough. In the *Consolations* he speaks
in favor of a "*refined plutonic view*," the refinement being
the elimination of heat as the formative cause of secondary
rocks and the rejection of a steady-state condition in the
earth's history.

> In the plutonic system, there is one simple and con-
> stant order assumed, which may be supposed eternal.
> The surface is constantly imagined to be disintegrated,
> destroyed, degraded, and washed into the bosom of the
> ocean by water, and as constantly consolidated, ele-
> vated, and regenerated by fire; and, the ruins of the
> old form the foundation of the new world. It is sup-
> posed that there are always the same types both of dead
> and living matter, that the remains of rocks, of vege-
> tables and animals of one age are found imbedded in
> rocks raised from the bottom of the ocean in another.[89]

But this is not what we find, he says.

> In a variety of climates, and in very distant parts of
> the globe, secondary strata of the same order are

found, and they contain always the same kind of organic
remains, which are entirely different from any of those
now afforded by beings belonging to the existing order
of things. The catastrophes which produced the
secondary strata and diluvian deposition could not have
been local and partial phenomena but must have extended
over the whole, or a great part of the surface, and in
the rocks which may be regarded as more recently deposited, these remains occur but rarely and with abundance of extinct species--there seems, as it were, a
gradual approach to the present system of things and a
succession of destructions and creations preparatory
to the existence of man.[90]

In these few pages Davy presents formidable arguments
against the steady-state earth of James Hutton, and offers an
outline of the evidence which favors an interpretation of a
purposeful progression toward higher life forms culminating
in recent times with the appearance of man. The publication of
this clear and concise expression of the progressionist view of
the earth's history was timely, for Charles Lyell, who quoted
extensive passages from the *Consolations* in the introductory
section of his *Principles of Geology,* wrote of Davy as "a late
distinguished writer" who had "advanced some of the weight-
iest . . . objections" to the uniformitarian doctrine of a
steady-state earth, objections that Lyell was compelled to
answer.[91]

Lyell's attempt to refute these views may be the best testi-
mony we can offer for Davy's contemporary reputation as a geologist. But because Davy always strived to be more than a geologist or a chemist (he commonly spoke of himself as a philosopher), he has provided in these geology lectures something more
than a geologist's insight into the state of that science as
it was in 1805.

Humphry Davy on Geology

Introductory Lecture
for the Courses
of 1805

The love of knowledge and of intellectual power is a faculty belonging to the human mind in every state of society; and it is one by which it is most justly characterised--one the most worthy of being cultivated and extended.

Useful to the individual, and even necessary to his existence, its general effects upon the species are, in the highest degree, important and beneficial; and the improvements in the condition and in the happiness of mankind appear, in all instances, to have preserved a uniform pace with the progress of the inventions of the arts and the advancement in the sciences.

This truth scarcely requires any demonstration. To prove it, there is no necessity for recurring to any refined arguments; the mere comparison of the rude and of the cultivated state of society must carry conviction to every unprejudiced understanding. In the dreams of a brilliant imagination, indeed, the uncivilized state of man may appear in high and vivid tints of happiness; and the fancy of an enthusiast may enable him to draw strong contrasts between nature and art unfavorable to the latter: between blue skies, verdant groves, murmuring streams, the scenery of a mountain country, and the smoke and dirt of towns, the noise and bustle of commerce, and the insipidity of productive plains; between the earth wildly and spontaneously producing food, and grounds made fertile only by human labour; between the vices, miseries, and dependence of man in society, and his simple virtues, his lofty pastoral manners, and his unsubdued freedom in the condition of nature.

Such romantic pictures, though they should be adorned with the highest colouring of genius, can, however, scarcely in the slightest degree affect the opinion of sound and judicious reasoners. The details of past times, the narrations of travellers,

or even a simple observation of the habits and propensities of
the human mind are sufficient to demonstrate that its highest enjoyments are connected with an active state of the understanding
and an exalted social intercourse--are sufficient to demonstrate
that the being whose pleasures are only produced by the gratification of his common wants, and whose wants are constantly limited
by the poverty of nature, can never be justly opposed to the man
whose delights are, in a great measure, conformable to his
wishes, whose intellectual gratifications are even more numerous
than his appetites, and whose mind is the *master* of his body and
not its *slave*.

     Many speculative men, whose minds have been awake to the ad-
vantages of improvement, have, nevertheless, conceived that, in
all cases, there must be certain limits to the progress of civil-
ization--have conceived that the sciences and the arts, however
beneficial in their first effects, must finally tend to enfeeble
the character and to promote the increase of luxury. Such persons have generally founded their opinions upon incorrect views
of the history of ancient nations and, haunted by ideas of their
rapid elevation and downfall, have believed that the same powers
operate in modern times, and that the germs of the ruin of states
exist in those very causes which have produced their greatness.
That power, riches, and leisure are essential to a great extension of philosophy and literature, and that they are likewise
often the causes of vice and depravation, cannot, indeed, be
denied; but a few facts, derived even from the history of the
empires of antiquity, will distinctly show that the influences
of the arts and sciences, in great and wealthy states, tend
rather to depress than to promote common luxury, and that those
periods the most distinguished by elevation of moral character,
by the social virtues, and by the higher feelings of the soul
were likewise the periods in which philosophy and letters were
most cultivated, and in which the fine arts were ardently pursued. The most happy period of Grecian civilization is that between the first Persian and the second Peloponnesian wars. It
was at the beginning of this period that literature and science
made their first progress in Greece; and, at the time that they
were studied with the greatest ardour, the patriotic spirit and
the heroical virtue of the people were revealed in their full
splendour. It was at the beginning of this period when Anaxagoras
was instructing the youth of Athens in speculative philosophy,
when Hippocrates was laying the foundation of medical science,
when Democritus was pursuing the paths of experimental enquiry.
It was at the same time that the mind of the inhabitants of
Attica were kindling with the poetic feeling raised by the immortal genius of Homer. It was at the same time that they prose-
cuted their most active war of liberty against the Persians, that
Miltiades and Themistocles led on their troops to conquest. It
was at the same time that Leonidas and his 300 Spartans fell at
Thermopylae, martyrs in the cause of freedom, and glorying that
they were permitted to die for their country. It was toward the
close of this period that sculpture, painting, and the arts of

life flourished in full vigour. It was at this time that the
chisel of Phidias raised out of the rude marble forms of majesty
and grace. It was at this time that the same dramatic poet,
Sophocles--whose immortal compositions will ever continue as
models of excellence--appeared as a warrior and conqueror at the
head of armies; and that the same philosopher, who was called
the wisest of men, endured all the hardships that the life of a
common soldier can offer. And, perhaps, the character of the
age can scarcely be better delineated than in one of the incidents in which he was the actor. Alcibiades, his disciple, was
wounded in the battle of Delium. Socrates carried him on his
shoulders, defended him, and dared, at the same moment, to expose
his life for his friend and his country.

Traits of a very different character marked the later
period of the republics of Greece; and, at the era when they
were about to resign their liberties to the power of Macedon,
the sciences and arts no longer flourished in their ancient
seats, but had passed into the country of the conqueror. At
this time Aristotle was obliged to fly from Athens, and a law
was passed to prevent any teacher of philosophy from opening a
school in that city, which had before been the theatre of her
glory. Luxury and sensuality only occupied the minds of the
people; and no persons were distinguished by public approbation
except such as gave great entertainments, or such as possessed
fortunes which enabled them to gamble largely. And as we are
informed by Athenaeus, the year before the invasion of Philip,
the freedom of the city (an honour which before had been bestowed only upon potentates, warriors, and philosophers) was
given to two young men, whose only merit was that their father
had been one of the best cooks in the commonwealth.

The same principles might be illustrated by examples from
many other nations. Rome affords as many instances, and of precisely the same kind as Greece. And at a later period, the elevation, progress, and the decline of the power of the Arabians
depended upon similar causes. The only time at which this last
people were truly great and happy was at the time in which literature and science were patronized by Almansor and his suc-
cessors, the caliphs of Baghdad. At the moment that the desire
of intellectual improvement disappeared, the savage and sensual
spirit of the religion of Mahomet carried its followers on to
ruin, even amidst the triumphs of conquest; and the progress of
the crescent, at first marked by victory and desolation, soon
finished in ignorance and debility. All experience, all analogy,
decidedly proves that unless power and riches are employed for
increasing the sources of mental gratification, and for keeping
alive the activity of the soul, their tendency must be evil;
and all the elder nations who have fallen from greatness, have
offered before their ruin similar characters: wealth without
science, improving manufactures or commerce; the few the conquerors of the many; and great cities peopled only with soldiers,
with rich men, with parasites, and slaves.

The refinements of the useful or ornamental arts in modern times bear no relation to the luxuries of the civilized nations of antiquity; and, as they are at present pursued, they are amongst the first causes of the general improvement of society, for they not only promote individual comfort, but they afford constant objects for employment: they preserve the love of invention; they promote emulation and the desire of excellence amongst the labourers in the same department; and they tend to unite the different classes of society by ties of usefulness, of mutual dependence, and of mutual advantage.

To the superficial observer, the attempt to extend the refinements of inventions beyond that state in which they are fitted for all the useful purposes of life, may appear wholly unnecessary; but it should be remembered that, in aiming at perfection in a manufacture, the workman is constantly improving himself; and, in attempting to produce articles which are to sell at a high price, he makes a number much better than they would otherwise be, which are disposed of at a moderate rate.

A finely polished knife, for instance, which costs a guinea, may not have a better edge than one which sells for a shilling only; but the cutler who has produced the expensive knife, from his accurate acquaintance with his art, gained from habit and laborious operation, is able to make the common knife better, and at a lower rate. A thousand cases of the same kind might be adduced. The elegant models of the Etruscan vases, produced by the ingenuity of the late excellent Mr. Wedgwood, may be said to have no immediate application to common uses; but yet, in consequence of their invention, a spirit of imitation and emulation has operated upon every branch of the porcelain manufacture, and even the forms and composition of our common pitchers and common flowerpots have, in consequence, been improved.

In certain departments of industry still greater advantages result from the constant attempts to attain the highest degrees of excellence. Examine agriculture. No person, who understands the luxuries of the table, will assert that a sheep, rendered enormously fat in rich pasture or on turnips, is better or a greater luxury than one that has grazed on the aromatic herbage of the Welch mountains--but in the attempt to produce this well-fed animal, which perhaps gains the prize at Smithfield, a number of others have been improved to a less extent, and rendered, in consequence, more adapted to common use. And the high price of well-fed cattle has awakened the feeling of emulation amongst farmers, in consequence of which the nature of the best breeds of cattle has been studied, the manner in which they can be most efficaciously nourished considered; and, from the extension of such enquiries, all the principles of farming have been more minutely investigated, and the art of cultivating land improved and adorned with new discoveries. The principle is general: whenever manufactures or any productions of art become articles of general consumption, the higher and more expensive refinements of them are absolutely necessary, not merely for their improvement, but likewise to prevent their decline.

All parts of the system of commerce are intimately connected.
Honourably acquired wealth in such a country as that in which we
have the happiness to live--honourably acquired wealth, I say--
produces credit, and from credit arises capital capable of an extent almost indefinite. Hence proceeds the division of labour;
hence the invention of machinery; hence the circulation of wealth
and power from one extremity of the empire to the other, and
communicating, like the vital blood flowing through the vessels,
to every part health and strength. Hence all the productions of
the globe are made subservient to the uses of man; and nature
arises subdued by artificial means, not impoverished or deformed,
but enriched, and rendered more beautiful.

The useful arts in modern times have attained an infinitely
higher degree of perfection than in the most splendid eras of
antiquity, and the improvements and extension of the sciences
will admit of no parallel instances. That light of knowledge,
which was only dimly perceived by the ancients, which was obscured by the clouds of error and of prejudice, has appeared to
us in all its purity and brightness; and whilst nature, and the
order established by the author of nature, have been to a great
extent developed, the science of man has not been neglected. The
works which awaken the imagination and exalt the feelings have
preserved all their effect upon the mind. By means of experiment
a new creation, as it were, of facts has appeared--of facts as
much superior to mere speculations as things can be to words.
Letters, the great instruments of thought, have assisted science,
and science has given new objects and new forms for adorning and
extending literature. All the different branches of knowledge
have assisted each other, and like different instruments of music,
the sounds of which combine in harmony, they have all cooperated
in enlightening the mind, in extending its enjoyments, and in
exalting the state of social life.

Though so magnificent a structure has been raised in science
rapidly and as if by a kind of enchantment, yet it is still un-
finished and new labours and new efforts of ingenuity are required both for ornamenting and extending it, and for preventing
any of its parts from falling into decay. Knowledge is like a
river which, unless its springs are constantly supplied, soon becomes exhausted, and ceases to flow on and to fertilize. The
mind requires novelty even as a stimulus to exertion; and the
philosopher who has made a discovery in natural science, or the
author of a work of genius in art or in literature is a bene-
factor, not only to the present generation, but likewise to future
ages; for he gratifies that faculty of enjoyment which is pure
and intellectual, and which must be more exalted as society becomes more improved.

Very few persons in the present day are disposed to reason
against the advantages resulting from the higher refinements of
science and philosophy, and the only argument that can be brought
forward is one founded upon the doctrine of common utility. When
a new fact, for instance, is ascertained in chemistry or in electricity, the superficial observer is very apt to slight it, if it

does not immediately admit of some application to the common
wants of life. This, however, is very unfair, for all experience
proves that the greatest and most important inventions which have
arisen from scientific principles have never been ascertained
till long after the principles themselves were developed; and so
intimately connected are all the objects of human enquiry, and
so much dependent upon the sensible properties of bodies, that
it is scarcely possible that any great theoretical improvement
can be made without being soon accompanied with practical advantages. A newly discovered country ought not to be neglected,
though it cannot be immediately brought into cultivation, because
it does not immediately produce corn, and wine, and oil.

But, independently of these considerations, all truths in
nature, all inventions by which they can be developed, are worthy
of our study, for their own sake, rather than with any idea of
profit or interest. Whatever can enlarge the views of the mind,
raise new sentiments of intellectual pleasure, or make us acquainted with new properties and powers in the substances surrounding us, is in the highest degree worthy of the pursuit of
a being whose noblest faculties are reason and the love of
knowledge.

All the discoveries, all the works of human genius, are of
great importance to the community; but that their full effects
may be produced, it is necessary that the public mind be prepared
to enjoy them and to estimate their advantages. The general diffusion of letters and philosophy is necessary to the progress of
the higher inventions of the mind; for unless the labours of men
of ingenuity meet with public support and approbation, they can
never be actively pursued, and must soon languish and die. All
minds require hope to animate them to exertion, and the desire of
glory is one of the most common to great and elevated understandings. The increase of general knowledge must uniformly produce
the general patronage of letters and philosophy, and this is a
most excellent and important end. Men of genius, in former
times, have often languished in obscurity, not because their
merits were neglected, but because they were not understood.
This, however, can scarcely happen in the present day, in which
all sources of useful information are laid open, and in which un-
paralleled exertions have been made in the higher classes of
society to diffuse improvement and to promote all objects of enquiry which can benefit or enlighten the public. There are other
uses, still greater uses, resulting from the communication of
general and popular science. By means of it vulgar errors and
common prejudices are constantly diminished. It offers new
topics for conversation and new interests in life. In solitude,
it affords subjects for contemplation and for an active exercise
of the understanding; and in cities, it assists the cause of religion and morality by preventing the increase of gross luxury
and indulgence in vicious dissipation. Man is designed for an
active being; and his spirit, ever restless, if not employed upon
worthy and dignified objects, will often rather engage in mean
and low pursuits than suffer the tedious and listless feelings

connected with indolence; and knowledge is no less necessary in
strengthening the mind than in preserving the purity of the affections and the heart.

Some few arguments are now and then brought forward against
the efficacy of popular instruction. It is urged that superficial and general knowledge often tends to produce pedantry, and
that persons who are only imperfectly learned are sometimes vain
and presumptuous. With regard to the charge of pedantry, it can
only be applied to the half-taught in manners, as well as in
science; and, in such a refined period as this in which we live,
it is scarcely possible that such a folly can flourish. What is
sometimes called pedantry, indeed, may depend upon the ignorance
of the many, as compared with the knowledge of the few; but the
moment the language of science becomes the common language of
refined society, every feeling of this kind must cease, and till
that event takes place, the person must be very, very deficient
in common sense who endeavours to astonish by a parade of knowledge, and who, being in possession of a light, chooses rather
to employ it for dazzling the eyes of others than to use it for
his own guidance.

That persons who are only *beginning* to attend to the principles of science often overrate their acquirements and abilities,
cannot be denied; but this is a circumstance of very little importance and seldom of much permanence. In every well-regulated
mind false confidence cannot be of any long duration. Vanity
almost always carries with it a certain cure. Disappointment
soon follows the ardent hopes of wild presumption, and in a
sound understanding, the conviction of having been once mistaken generally produces discretion and caution, which daily
become more habitual, which direct the mind in its judgments,
and which, when combined with feeling, become the foundation of a
just and accurate taste.

All human knowledge is necessarily imperfect; but the
further it extends, the better are its effects. No efforts
made for the attainment of truth ought to be slighted. Lofty
ideas are often connected by man even with his weakness and
follies: how much more ought they to arise from his strength and
his wisdom! His powers are often wasted in attempts to obtain
trifles which vanish or cease to delight at the moment they are
in his possession, and we ought always to rejoice when those
powers are applied to objects which are permanent and connected
with true glory. Man is formed for pure enjoyments; his duties
are high, his destination is lofty; and he must, then, be most
accused of ignorance and folly when he grovels in the dust, having wings which can carry him to the skies.

Lecture One

So different are the exertions of the faculties of the mind, and so infinitely various the combinations of our ideas, that the same objects may be examined with the most opposite views and considered under many diversified and beautiful relations. It is on this fact of our nature, so familiar to every understanding, that the great extent and progression of science and philosophy depend. Hence their division into various branches and hence the distinctness and accuracy of the different species of knowledge.

The planet that we inhabit may be considered in its connection with the general system of the universe, as acted upon by gravitation and revolving around the sun. It may be examined as the abode of organization and life, covered with vegetation and peopled with animals. Or it may be studied as composed of different inorganic parts variously arranged and subservient to different uses. It is under this last point of view that it is the subject of geology. The word is derived from the Greek language and signifies the science of the earth, but its acceptation is limited, and it is only applied to the branch of knowledge relating to the nature, position, and changes of the bodies that compose the known part of the surface of the globe.

The facts belonging to geology are numerous. The objects of the study are as many as the substances found in our soils, our rocks, and our mountains, in the sea, and in the atmosphere. And the alterations that the great parts of our system are continually undergoing can only be explained on the principles of this department of philosophy.

The information to be derived from geology is applicable to many purposes. In a number of instances it may be made sub-

10

servient to the wants of life. And all the theoretical views
belonging to it, when accurately pursued, lead to beautiful
truths.

Fixed on the earth and dependent for our existence upon the
various objects surrounding us, many of our necessities are supplied and some of our comforts produced by applications of the inorganic substances presented to us in nature. The soils from
which we raise our vegetable nourishment, the stones of which
our habitations are formed, the fuel we employ for so many purposes, and the metals absolutely essential to civilized man are
all subjects for the application of geology. And it is not
merely their nature which is indicated by this science, but
methods of discovering them and their extensive relations.

The rocks and stony substances which are found beneath the
soil and which compose the foundations of our hills, though so
various in their appearances, yet contain many common substances.
The same bodies are found in very distant countries. There is an
order in their position, and the general nature of the mineral
bodies belonging to a particular district is in most cases indicated by the examination of a few specimens.

In a mountainous country, for instance, if a stone of this
nature occurs (which is granite), there is great reason to suppose that this substance will be likewise found at no great distance.[1] But in such a position it would be vain to search for
shell limestone or for coal.

Again when shell limestone is found, there is a strong presumption that soft-black slate or loose sandstone will likewise
occur in some contiguous spot; and in such a country there can
be little doubt that at some depth or in some direction pit coal
would exist.

These circumstances, which are fundamental facts of the
science, prove in a strong manner that considerable discoveries
may often result from geological knowledge. And it affords information capable of being immediately brought into use. The
person who is digging for pit coal, when he meets beneath the
soil granite rock, if acquainted with the arrangement of and nature of strata, will be immediately instructed to give over his
labour and to spare useless expense. But should he find sandstone, a substance which to an uninstructed eye appears much of
the same nature as granite, it affords him immediate encourage-
ment to proceed in his researches. And a yellow or red sandstone
on a soft slate bearing the impression of vegetable leaves would
offer indications almost certain of the substances sought for,
for these rocks are generally immediately incumbent on coal.

Similar reasoning may be applied to metallic veins. The
metals seldom or never occur in rocks of serpentine or syenite,
or in soft coaly schist,[2] nor in sandstone, nor in basalt--but
they may be looked for in soft granite, in hard schist, and in
hard shell limestone. And if in granite or schist a vein of
white stone is found remaining in a direction from east to west,
there is much probability that in some part of its depth it may
afford useful metal.[3] And if veins of spar[4] occur in rocks

partly hollow and partly filled with a yellow substance of this
kind, which in Cornwall is called gossan,[5] it may be always concluded that these veins will be productive, and the larger the
quantity of gossan the better the indication. A number of the
same kind of instances might be adduced.

The study of the science is not useful merely to the miner,
for as it explains the general direction of strata and the substances they contain, it offers application to many more of the
arts of life. It ought to be studied by the engineer who is employed in the construction of canals, as certain strata exceedingly hard are often [found] with others very soft and easily
cut through. The drainer in order to make his operations successful ought to be minutely acquainted with the arrangement of
the rocks in the district from which the springs arise that it
is his business to divert. And the improver of land may often
derive from geology very useful instruction with regard to limestone, marl, clays, their application, and the nearest places
whence they may be procured. It is according to these views
that the study may be of national as well as of particular advantage.

Such reasons show that it is highly worthy of being pursued, and there are other reasons of no less importance, though
less obvious and more refined, which ought not to be slighted in
treating of its relations and general influence. These reasons
will appear indeed perhaps more precisely to general enquirers
after information, for they depend upon circumstances which can
scarcely fail to be equally interesting to every refined understanding.

It will be needless to enter upon a subject which has been
often discharged in this room, the general usefulness of the
knowledge of nature in increasing mental enjoyment and in
strengthening and exalting our sentiments. But it surely will
not be deemed improper to point out some of the peculiar intellectual effects of the objects of these lectures and some of the
advantages of the study as compared with that of other branches
of science. Of all the material objects which can employ our
attention, those which are nearest to us ought to excite the
warmest and the most immediate interest. And after man and animated nature, no subject of physical enquiry bears a more distinct relation to us than the place of our abode--the earth--
which is the cradle of our existence, which affords the instruments of our power, and which, though not the great parent, is
the peaceful and useful nurse of some of the most constant and
uniform of our treasures.

The study of the constitution of the globe, of the manner
in which its changes are produced, and of the laws by which dead
and inorganic matter are rendered subservient to the purposes of
living beings, affords some of the most sublime objects which
even the most insensible mind can scarcely consider without deep
feeling. And they are capable of filling even the most comprehensive understanding and of calling forth the highest exertions
of genius.

The rocks which are elevated in the clouds or those which
resist the waves of the sea, their composition, the laws of
their formation, and their alterations in consequence of the action of the atmosphere and of water, these form only a part of
the subjects of discussion of geology. For this science equally
relates to the nature of soils, of all the solid strata of our
earth, to the gradual changes produced by the ocean upon the
land, and to the great convulsions of nature occasioned by inundations, by volcanoes, and by earthquakes.

The attainment of the knowledge belonging to these highly
interesting subjects is founded almost wholly upon observation,
for in a few cases only can instruments of experiment be applied.
And the acquaintance with mineral bodies, which constitutes one
of the most laborious and most interesting of the study, may in
almost every case be derived from an examination of their external character.

Geology, perhaps more than any other department of natural
philosophy, is a science of contemplation. It requires no experience or complicated apparatus, no minute processes upon the
unknown properties of matter. It demands only an enquiring mind
and senses alive to the facts almost everywhere presented in
nature. And as it may be acquired without much difficulty, so
it may be improved without much painful exertion.

The person who is devoted to geological enquiries can
scarcely ever want objects of employment and of interest. The
ground on which he treads, the country that surrounds him, and
even the rocks and stones removed from their natural position
by art are all capable of affording him some degree of amusement; and every new mine or quarry that is opened, every new
surface of the earth that is laid bare, and every new country
that is discovered offers to him novel sources of information.
In travelling he is attached to a pursuit which must constantly
preserve his mind awake to the scenes presented to it. And the
beauty, the majesty, and sublimity of the great forms of nature
have their effect in the imagination rather increased than diminished by being connected with the views of philosophy.

The imagery of a mountain country, which is the very
theatre of the science, is in almost all cases highly impressive
and delightful, but a new and a higher species of enjoyment
arises in the mind when the arrangements in it, their harmony,
and its subserviency to the purposes of life are considered.
To the geological enquirer every mountain chain offers striking
monuments of the great alterations that the globe has undergone.
The most sublime speculations are awakened, the present is dis-
regarded, past ages crowd upon the fancy, and the mind is lost
in admiration of the designs of that great power who has established order in which at first view appears as confusion.

It will not, I trust, be necessary to pursue these views to
any greater extent, for sufficient, I hope, has been said to
prove that the study is adapted to almost every mind and that
any labour which may be bestowed upon it can scarcely be considered as vain.

A very few words will be sufficient to explain the plan
which it is my intention to adopt in developing the principles
of the science and in rendering them plain and intelligible. As
it may be supposed that some of my hearers are unacquainted with
many of the terms belonging to geology, and as most of these
terms will appear barbarous and are really difficult to be remembered, I shall introduce them as gradually as possible, beginning with the history of science.

In this history the early opinions of the ancients concerning the elements of the globe and their changes will first be
discussed, and the more refined theories of the moderns which in-
volve the outline of the science will be afterwards considered.
In the general details the opinions that are complicated will
follow those that are simple and distinct.

After the history of the science the general order of the
facts belonging to it will be explained: the nature of rocks
and stones, the relations of the different strata, and the
changes that they are at present undergoing. This part of the
subject I shall illustrate as much as possible by large specimens, by drawings, and by picturesque sketches of the position
of rocks. From this last method, much useful information, it is
hoped, may be derived.

This painting will show the nature of the plan to be
adopted.[6]

From the liberality of the managers I shall be able to
exhibit a number of the same kind and on the same scale, and
most of them will contain accurate portraits which will be the
more interesting as belonging to the island we inhabit.

After having examined the existing state of the globe,
the last subject of the course will be the discussion of the
theories that have been formed to explain the nature and position of its parts, the laws of their formation, the changes that
they have undergone in the lapse of ages, and the principles and
orders by which the whole is continually preserved. It is this
department of the science which will perhaps be the most interesting. It is the one that offers the greatest objects of contemplation. It has most employed human ingenuity, and many of
the theories and speculations upon it exhibit brilliant examples
of the most powerful efforts of genius.

The task which I have proposed to myself in explaining
all those various objects is one which in this country, I believe, has never been before undertaken, and it is one connected
with much labour and with many difficulties. As yet no elementary treatise has been published on geology. I have been obliged to form every part of the arrangement and to collect the
different facts from very distant and very scattered sources.
It is a subject in which very few only of its parts will admit
of the arrangement of language, of beauty of description, of
elegance of expression. I shall spare no exertions to make it
useful and intelligible. In mentioning these circumstances I
mean to apologize only for imperfections that cannot be
avoided.

The history of geology is a subject that necessarily leads us to consider the state of the human mind even in the elder periods of time. Man in his most savage [state] must have been strongly impressed with the forms, appearances, and changes of the earth and its diversified objects. And some of his earliest reasonings directed from effects to causes must have been con- nected with speculations on unknown times, on unperceived alter- ations of matter, and on the wonders of creation. In all the earliest systems of religion of the heathen nations that have come down to the present time either by history or tradition, certain doctrines are uniformly found relating to the formation of the earth and its early changes. And most of these doctrines appear to have been in some manner derived from the grand account of the creation and of the deluge delivered in the sacred writings.

The cosmogony of the Hindus offers a remarkable instance and is quoted as such by Sir William Jones from the account of the formation of the universe supposed to be given by Manu, the Son of Brahma, to the Indian sages.

> The world, says he, was all darkness, undiscernible,
> undistinguishable, altogether, as in profound sleep,
> 'till the self-existing invisible God making it manifest
> with five elements and other glorious forms perfectly
> dispelled the gloom. He, desiring to raise up various
> creatures by an emanation from his own glory, first
> created the waters and impressed them with a power of
> motion; by that power was produced a golden Egg blazing
> like a thousand Suns on which was born Brahma, self-
> existing, the greatest parent of all rational beings.[7]

To an unprejudiced enquirer this passage must appear as a para- phrase of the more simple and more sublime detail of the in- spired writer,

> In the beginning God created the heavens and the Earth
> and the Earth was without form, and void and darkness
> was upon the face of the deep, and the spirit of God
> moved upon the face of the waters and God said, let
> there be light, and light was.

The early Egyptian cosmogony, according to a late ingenious writer Captain Wilford, is very similar to that of the Hindu and is probably derived from the same source, though the account of it is less distinct. [Keb],[8] or the supreme being, is supposed in their mythology to be the one author of being and the father of Osiris or the sun, the object of adoration who is imaged as issuing in the form of light from an egg, the first of created beings.

Amongst nations disposed to polytheism the great tradition and primary doctrine must soon have been blended with a number of opinions and dogmas arising from the condition and peculiar

circumstances of society. And new reasonings concerning the
universe, its productions, the elements of matter, and the
causes of things must have been constantly awakened in the mind.

To common understandings alive to the impressions of superstition, it is easy to conceive that all natural objects connected with strong feeling would readily be referred to some unknown divine power, and hence the different parts of nature be
represented in an extensive mythology. And facts prove that in
nations the most distant, this has actually taken place and in a
similar manner. Thus the winds have been personified. Fire,
one of the most beneficial as well as one of the most powerful
natural agents, has been worshipped under different symbols in
almost every part of the ancient world. And that power supposed
to be individual which raises water from the ocean and causes
it to fall upon the land, giving vigour to the plant and the
animal, has been equally adored in India, Egypt, and Greece: in
India as Iswara and Siva; in Egypt as Isis; and in Greece as
the celestial Venus, who is represented as rising from the bosom
of the ocean clothed in loveliness, the parent of beauty and the
preserver of life.

Long before men had examined the present, they endeavoured
to reason concerning the past and to predict concerning the future. The imagination is always more active than the reason in
an early state of society. The first person who professed to
teach others attempted to combine the character of the poet, of
the philosopher, and of the priest; and the only profane records
of the opinions of early ages that have been transmitted to the
present time are adorned with the language of passion and
cloaked with the graces of the fancy.

The Greeks have represented the bards as their first in-
structors. The names of Linus, of Orpheus, and of Museus are
celebrated by their earliest historians. And the hymns of
Orpheus, if they be admitted to be genuine, contain some general
expressions of facts with regard to nature and the powers acting
in nature which are very admirable and which could scarcely be
expected in so early an age. His "Hymn to Jupiter" is filled
with beautiful poetry and with some comprehensive truths concerning the production of things and the cause of their relations and
order. I have made a free translation of it which I shall venture to read. The original, even if its high antiquity be
doubted, will at least be found impressive from its views.

> Almighty Jove, the first, the last,
> The present, future, and the past
> Source of all existing things,
> From thee created being springs.
> From thee the heavens derive their birth,
> From thee arose the solid Earth,
> From thee the radiant solar light,
> From thee the stars that gild the night.
> Thine is the morning changeful hue,
> Thine is the ether bright and blue,

> Thine is the warmth of summer days,
> Thine is the land, the air, the seas.
> Thy substance is our vital breath,
> Thy spirit raises life from death.

In the *Theogony* of Hesiod, as has been observed by the great Bacon, there is evidently much allegorical description which may be supposed to apply to the constitution and changes of the earth. But so many explanations of the same fables may be given that any attempt at accurate deciphering would be wholly vain. It would be wandering in a labyrinth without end, not for the sake of discovering truth, but merely with a view of detecting error.

The poems of Homer, evidently composed at a time when the dawn of civilization was opening upon the human mind, contain infinitely more facts relating to man than to nature, and however great the extent to which mythology had been applied before his time in explaining the phenomena of the universe, it is chiefly made in the *Iliad* and the *Odyssey* the machinery for exciting the interest and awakening attention to the actions of heroes and of men. Occasionally, indeed, the physical attributes of the gods occur. And Jupiter is represented as the power that acts in the air, Neptune as the moving force of the waters, and Pluto as the principle residing in the solid earth. Thus in the battle in the twentieth book:

> Above the fire of Gods his thunder rolls,
> And peals and peals redoubled rend the poles.
> Beneath, stern Neptune shakes the solid ground:
> The forests wave, the mountains nod around,
> From all their summits tremble Ida's woods
> And from their sources boil her hundred floods.
> Troy's turrets totter on the rocking plain
> And the tossed navies beat the heavy main.
> Deep in the dismal region of the dead
> The infernal monarch reared his horned head,
> Sprung from his throne lest Neptune's arm should lay
> His dark dominions open to the day.

The first accurate accounts of the general philosophical opinions of the elder nations concerning the universe freed from the mysteries of mythology and from poetic fables are those delivered to us in the history of the philosophical schools of Greece. At the era, however, of the foundation of the Grecian republics, science was far from being indigenous in that happy country, and both the legislators and philosophers were obliged to travel into foreign lands in quest of information.

Egypt, as we are instructed by Herodotus and by various other authorities, was the great nurse of Grecian science and the opinions of the first European sages are said to have been modifications of the doctrine of the Egyptian priests. Of these doctrines in their pure and esoteric form very few documents re-

main and by the fertile imagination of the Greeks they were soon
blended with their own opinions. Water, it is said, they considered as the great active element, the cause of the changes of
the globe and of the reproduction of life. Fire, likewise they
regarded as a principle, and air as a modification of water and
fire.

In the *Timaeus* of Plato the opinions of an Egyptian sage
are detailed as it is supposed they were delivered to Solon at
Sais. From them Plato acknowledges that he borrowed the idea of
the Atlantis fable, and they are so very singular and contain so
many opinions different from those of the Greeks concerning nature and the changes of the globe that I shall give a translation of the most interesting part.

> You Greeks, says the Egyptian philosopher to Solon,
> you Greeks are always children, nor is there an ex-
> perienced man amongst you. All your knowledge is
> lately acquired, neither containing any ancient opin-
> ion derived from remote tradition, nor any truth of
> ages that are long past.
>
> And the reason of this is the multitude and variety
> of the destructions of the human race which have been
> and which will be. The greatest of these indeed aris-
> ing from fire and water, but the lesser from ten thou-
> sand other contingencies. That Phaeton, the offspring
> of the sun, attempting to drive the chariot of his
> father and not being able to keep the track pointed out
> to him burnt up the substances belonging to the earth
> and perished himself blasted by thunder, is indeed con-
> sidered as fabulous. Yet it is in reality true. For
> it expresses the mutations of the bodies revolving in
> the heavens above the earth, and indicates that after
> long periods of time a destruction of terrestial nature
> ensued from the devastation of fire. Hence those who
> dwell either on mountains or in lofty and dry places
> perish more abundantly than such as dwell near rivers
> or on the borders of the sea. To us indeed the Nile
> is both salutary in this respect and in others, and it
> preserves us from such devastations.
>
> Again when the gods, purifying the earth by water,
> deluge its surface, then the herdsmen and shepherds in-
> habiting the mountains are preserved while the people
> of your cities are hurried away to the sea by the im-
> petuous inundations of rivers. On the contrary in our
> region neither then nor at any other time did the
> waters descending from above pour with desolation on
> the plains, but they naturally rise upwards from the
> bosom of the earth.
>
> And from these causes the most ancient traditions
> are preserved in our country. Indeed it may be truly
> asserted that in the places where neither intense cold
> nor immoderate heat prevails, the race of mankind is

always preserved, though sometimes the number of in-
dividuals is increased and sometimes diminished. But
whatever has been transacted either by us, or by you, or
in any other place, whether beautiful or grand or wonderful, is to be found in our temples preserved to the
present day. While on the contrary you and other nations commit only recent transactions to writing and
the other inventions which society has employed for
transmitting information.

And since when at stated periods of time these
changes ordained by heaven, as if diseases of nature,
destroy them among you, you who survive are both desti-
tute of literary acquisitions and the inspiration of
the muses. Hence it happens that you become again as
it were an infant people and ignorant of the events
which happened in ancient times whether in foreign lands
or in your own country.

Of the Grecian philosophers, Thales the Miletian is the
first of whose opinions concerning the earth we have any authentic documents. He is said by Apollodorus to have flourished
about 600 years before Christ. His principal physical doctrine
was that which has been before mentioned as derived from Egypt,
that water was the principle of all things, and that being set
in motion by the supreme cause, the earth and all the forms of
being proceeded from it.

Anaximander, the countryman and disciple of Thales, as
there is every reason to believe, went far beyond his master in
the study of nature. He considered an infinite cause as the
creator and mover of matter. He supposed that though the parts
of the world are continually changing the whole is nevertheless
immutable. He developed, if we may believe Plutarch, what are
now known to be the true causes of the winds, that is to say, a
motion and change of place in the air produced in consequence of
the rarefaction of its lower parts by the solar heat.

Some of the opinions of Anaximander concerning the great
bodies of the universe are exceedingly curious. He is said to
have advanced that the sun is 28 times as great as the earth.
In his age such an opinion was exceedingly extraordinary, but we
should give him more credit for it if he had not said that the
moon was 29 times as large.

He conceived that the sun itself was not luminous but that
its light proceeded from a hollow circle surrounding it. This
idea might at first view be conceived to be very similar to that
of a celebrated modern philosopher, but the similarity will disappear when it is stated that the Greek considered the eclipses
of the sun as owing to the closing up of the orifices in the
circle necessary for emitting light.

The doctrines of Anaximander were taught after his death
by his scholar Anaximenes, who made many additions and ventured
to contradict generally the principles of Thales. He asserted
that air, and not water, is the chief material element, and that

water is capable of being composed from air, adducing as a proof
that by the condensation of air, rain is produced which when
congested becomes snow or hail, appearing then, as it were, a
species of earth.

But the greatest of the philosophers of the early school
was Anaxagoras who flourished about 450 years before Christ.
Born at Clazomene, he soon removed to Athens where he taught
philosophy for thirty years till he was banished from this
celebrated city from the mere circumstance of his asserting
that the sun was material and a mass of ignited substance.

Anaxagoras taught that one supreme intelligence was the
creator and governor of the world, and hence he was honoured
with the title of *Nous* or, the mind. Thus Timon speaks of him
in verses quoted by Laertius which I have thus translated.

> He spoke of one eternal living mind
> That all this glorious universe designed,
> And from compassion made it rise and shine
> In order and in harmony divine.

One of the most novel and most remarkable of the doctrines
of Anaxagoras was that he considered all bodies as composed of
similar parts or essences, and that if the ultimate figure of
particles could be obtained they would be found precisely similar to the original body which they composed.

This philosopher supported the idea of three elements--
earth, water, and heat. [A] few [of his ideas] are worth quot-
ing. The most remarkable of them perhaps is his speculation
concerning the nature of the heavenly bodies, all of which he
recorded as composed of materials similar to those of the earth.
And he endeavoured to prove his operation by stating that at
different times stones had fallen from the heavens. This last
idea is attributed to him by Aristotle, by Pliny, and by
Plutarch. Aristotle ridicules the notion, for in those days,
as well as in the present times, there were sceptics concerning
the bodies believed to have fallen from above.[9]

But Plutarch positively asserts, on the authority of
Daimachus, that in the second year of the seventy-eighth
Olympiad a large body of fire, which emitted many flashes like
shooting stars, was seen to fall near the river Egos in Thrace,
and that it left in its place a large and dark coloured stone.
[And Plutarch asserts on the same authority] that such an event
(and others of the same kind) would happen, was foretold by
Anaxagoras.

The few facts that we are in possession of with regard to
the first school of philosophy in Greece prove that as far as
their means extended, accurate observation was a guide to their
researches, but their views were necessarily limited. They were
unassisted by experiment, and instead of being rewarded for their
most enlightened opinions, they were constantly in danger of
persecution.

The people of that period were wholly unprepared to receive the truths of natural philosophy which opposed received
superstitions. In consequence, as will be shown in the next
lecture, the infant game of the science of observation soon
faded and perished, and then, as if dazzled by the light afforded
to them by nature, they soon closed their eyes and found it more
expedient, more profitable, and more conformable to the taste
of the age to amuse themselves with the visions of their own
creation.

Lecture Two

The number of Grecian philosophers and the variety of opinions attributed to them even in the first ages of civilzation of the people demonstrated in a striking manner the bias of public opinion towards study and the zeal with which science and information were sought for and pursued. The teachers of those times, however, in attempting universal systems formed a whole of which all the parts were necessarily imperfect. And the disciples, either following implicitly or altering only in a small degree the dogmas of their masters, were constantly diffusing theories in which truths and errors, speculations and facts were indiscriminately blended.

This was equally the case in all their departments of human enquiry, if the records of the early doctrines brought down to the present times be supposed authentic. Filled with the sacred strength of genius the first sages seemed uniformly to have believed that the human mind was capable of embracing all objects of knowledge. They delivered their opinions with the same confidence upon nature and man, upon the divine, and upon the human intelligence. And they seemed to suppose that the paths of practical as well as of abstracted sciences were to be developed by the mere powers of the unassisted understanding. But though all prone to hasty speculation in forming connections, yet many admitted of appeal to experience and observation. And it is to this circumstance, and to this alone, that we owe whatever is found true or valuable in their reasonings on sensible objects, on nature, and on the earth.

That the school of Thales had paid some attention to natural phenomena is evident from the accounts given in the last lecture. And one of the fundamental axioms of the fol-

lowers of Pythagoras was that the external signs of things were
no less necessary to the discovery of truth than reason and
analogy. The general doctrines of the Pythagoreans concerning
the appearances of nature had much more influence on the early
progress of science than any other system, and the greatest instruction amongst the ancients. And men whose opinions have
come down to the present time most unimpaired, and with the
greatest glory, belonged to this celebrated sect of philosophers.

The history of their illustrious founder is obscured to
so great an extent with fable and the marvellous that it is
very difficult indeed to trace any of the real occurences of
his life. Even his native place is unknown. Seven cities
only contended for the honour of having given birth to Homer,
but as appears from the testimonies of Diogenes Laërtius, Lycus,
and Josepheus, no less than nine different nations considered
Pythagoras as their countryman. He was, however, generally supposed to be of Samos, and his school is universally allowed to
have been established about 560 years before Christ at Crotona.

Like other ancient philosophers he travelled in quest
of information, and it is said that he derived a considerable
part of his knowledge from the priests of Egypt. He at least
learnt from them the art of professing one doctrine to the
public and of teaching another to his friends, a refinement
which though it satisfied the people for many years did not in
the end preserve him from persecution.

Pythagoras, if we may believe Plutarch, thought much,
talked little, and wrote nothing. Laërtius mentions, however,
the titles of some books supposed to be written by him, but
upon very loose authorities.

Iamblichus and Porphyry attribute to him wonderful powers
over nature and the faculty of predicting future events. Their
details, however, are not to be found in any of the more ancient
writers and appear to be forgeries; and the greatest wonder
faithfully recorded of him is that he was able to keep his disciples silent for five years of probation.

The doctrines of Pythagoras concerning nature, as recorded
by his followers, mark in the strongest manner boldness of conception, extent of knowledge, and possession of a mind filled
with some of the most important truths of mathematical science.
He endeavoured to apply the principles of number and quantity
to the explanation of the physical forms of bodies and the laws
of their changes. He considered matter as composed of certain
different primary elements. He made their number five and
assigned to each the form of a regular solid. The elementary
matter of earth he regarded as consisting of cubes. The principle of water he imagined to have twenty sides. Air he regarded as consisting of eight-sided figures, ether of twelve-
sided figures, and fire of pyramids.

Having discovered the monochord, he was so much delighted
with the invention as to search in every department of science
for analogies to harmony. And hence he conceived that the elements of bodies were united in proportions which may be ex-

pressed by the relations of numbers upon the musical scale, and
that all the infinite varieties of matter were deducible from
different orderly mixtures and combinations of the five geometrical elements.

This theory of the composition of the globe, though
founded upon assumptions which later experience has demonstrated
to be visionary, is nevertheless exceedingly ingenious and very
remarkable on account of the strong analogy it bears to the
system of the moderns concerning crystallizations. Romé de
l'Isle [1736-1790; DSB, 11:520] and the Abbé Haüy [1743-1822;
DSB, 6:178],[1] in their late refined speculations upon the forms
of minerals, have supposed six primitive or geometrical elements
capable, like those of the philosopher of Crotona, of being
united in various arrangements and of being again resolved into
their original forms.

The few features that are developed to us of the understanding of Pythagoras in an age of darkness and of ignorance
are at least sufficient to explain his great influence upon
philosophy. His discoveries, even as imperfectly displayed, are
of a high stamp, and it is not the least admirable trait of his
character that he supposed the idea of the roundness of the
earth and its revolution round the sun.

The first Pythagorean philosopher whose works are still
in existence is Ocellus the Lucanian who is supposed to have
flourished about 500 years before Christ in the beginning of the
celebrated era of the triumphs of Grecian liberty. His treatise
upon the universe--there is no reason to doubt its authenticity--[2]
is in the highest degree interesting as being the earliest book
composed upon a subject of pure science.

Ocellus, like his master, considered the earth as composed
of different regular parts, which, imperceptible by our senses,
are in a continued state of motion and of change. In his system
the solar fire is made the great agent in producing new arrangements of the bodies belonging to the globe. This substance he
regarded as the bond of union between all other elements, which
unites earth, water, air, and others. From the successive
union and separation of the elements he explains the order of
nature in which, though the parts are in continual change,
the whole remains one and unalterable.

Heraclitus of Ephesus and Parmenides of Elea, who flourished
about the time of Ocellus, professed a similar doctrine.
Heraclitus, not contented with supposing that heat was the great
active principle, held that it was convertible into all forms of
matter and that the pyramid by its modification produced the
other four regular forms of Pythagoras.

The high importance assigned in the ancient systems to
fire is not, indeed, to be wondered at when the general appearances of nature are considered. The sun, the source of the
vivifying warmth belonging to our globe, is likewise the cause
of the most numerous and most delightful of our sensations,
those of vision. By means of it water is raised from the ocean
and diffused through the air; by means of it vegetation is pro-

duced and the whole of the surface of the earth clothed with
verdure and ornamented with beauty.

The contrast of winter with spring, of night with day,
of the ice-clad mountains and sterile snow plains of the north
with the warm and fertile valleys of the south must have affected, in the strongest degree, the minds of those whose imagin-
ations are constantly alive. And the reasons and impressions
similar to these which have impelled so many uncultivated nations to offer divine honours to the principle of fire and to
exalt it into a god seemed to have operated upon the early
philosophers in inducing them to consider it as the most active
and energetic of the elements and as the material agent the
most connected with the changes of things, with motion, organization, and life.

A first intelligent cause of the order of the universe
was admitted by Pythagoras and his early followers; but many
of them, and particularly Ocellus, considered matter as having
constantly existed in form similar to its present arrangement.
And Timaeus the Locrian was the first philosopher of the sect
who supported the sublime and sacred doctrine of a primary
creator.[3] One of his treatises is at present in existence.
It is entitled *The Soul of the Universe*, and though obscured
with mysticism and in many parts unintelligible, it nevertheless contains some sublime and striking passages.

Timaeus considered all ideas of the forms of things as
preexisting in the supreme mind, and matter as modelled after
those ideas in order and arrangement in consequence of the
divine will. Earth he supposed to have been first produced,
then water, then air, and last of all, fire. From their mixture in the harmonious proportions of numbers, he considered
that the universe was created in visible and tangible form, perfect in its essence and indestructible as a whole, the temple
of the creator having all its parts adapted to the great ends
of life, and everywhere filled with animation, with sentiment,
and with happiness.

Timaeus is said to have taught philosophy in the middle of
the fifth century before the Christian era and to have been contemporary with Empedocles the Sicilian who was likewise a zeal-
ous Pythagorean. Some fragments of a poem of Empedocles's, *On
Nature*, are preserved by Diogenes Laërtius, from which it may be
discovered that his opinions concerning the origin of things
were similar to those of Ocellus and Parmenides.

He was a firm believer in the transmigration of souls, and
in two very ridiculous lines positively declares that he recol-
lected many of the former states of existence which he had passed
through. Amongst them he mentions those of a plant, a bird, and
a fish. He believed that he had lived in the earth, the air,
and the water; and to experience the effect of a new element,
fire, it is reported that he threw himself into the crater of
Mount Etna.

The age of the greatest political activity of the Grecian
states was likewise the age in which philosophy was examined and

extended with the greatest ardour. And an emulation similar to
that which animated the celebrated generals and warriors of the
time influenced the sages and induced them to exert the full
powers of their minds.

The Pythagorean system, though it possessed a great number
of admirers, was not without very powerful opponents, and Zeno of
Elea and Leucippus, said to be of Abdera, attacked the fundamental principles with great zeal. The physical opinions of
Leucippus only are preserved, and they are said to have been
similar to those of his master Zeno.

Leucippus, according to Diogenes Laërtius, conceived of all
things as formed of minute parts or atoms infinitely various in
their figures and without regularity, but constantly in motion
and endowed with a power of gravity or attraction, inducing them
to assume the forms of material bodies. From the attractive
powers and their primary motions he supposed the changes of
things to arise. And on these foundations, with the assistance
of the word "necessity," or "chance," he established his general
system, of which almost all the particulars are equally absurd
with the primary principles, for he held that the stars were a
collection of atoms kindled by their motion in a vacuum, and that
they are nearer to us than the sun which he supposed to revolve
around the moon. He considered the earth as flat and the sea as
constantly diminishing, and he vainly predicted that at some time
the whole surface would be dry land.

The only immediate follower of any reputation was likewise
of Abdera. He lived about 400 years before Christ and was
reputed to be an observer and an experimenter.[4]

The general doctrines of this school had very little cur-
rency. The common sense of men revolted at such opinions as
those which supposed harmony capable of arising without any
cause from confusion and intelligence as the result of a fortuitous mixture of senseless particles of matter. Even in minds
uninformed by true religion, the feeling of design and of a great
and useful order in nature is seldom absent. And however clouded
by superstition, it is still cherished; and the light, though dim
and feeble, is preferred to perfect darkness.

To enter upon a description of the minute differences between the disciples of the various schools would be an unpleasant
as well as useless labour. From the time of Anaxagoras, words
had been multiplying, though few new observations had been made.
And whilst theories of cosmogony were almost infinitely varied,
no accurate knowledge concerning the materials of the globe
had been obtained. Almost every student thought he could explain what the globe was in remote antiquity and what it would
be in future ages; but no one examined its present state and the
nature of its different parts. Every disciple of the Pythagorean
school confidently believed that he knew what matter composed the
sun, moon, stars, and the vault of heaven, whilst he was perfectly ignorant of the nature of the soils and the stones beneath his
feet. It appeared to all the philosophers more easy to imagine
than to observe, and more amusing to dream than to reason.

Socrates, the great Socrates, seems to have been the first person who had the good sense and the boldness to deride the systems of his age. His profession that in wisdom he felt this only, that he knew nothing, forms a striking contrast to the arrogant assertions of some of his contemporaries. No theory of Socrates concerning nature is delivered down to us, but there is every reason to believe that this most excellent man saw all the imperfections of the methods of the ancients and the vanity of their researches.

That he was devoted to experiment is evident from many testimonies. One of the great accusations against him in his most unjust trial, as quoted by Plato in his apology, was that he examined with unholy curiosity into things beneath the earth and in the skies. And it is very remarkable that the ridicule directed against him in the comedy of Aristophanes, *The Cloud*, is chiefly connected with his researches into the causes of natural things. This production affords us almost the only remains of his opinions on natural philosophy. He is represented by Aristophanes as attributing thunder and lightning to material influences existing in the clouds and not to the immediate anger of Jupiter, and as supposing that they are the cause of rain, of mist and snow, and that the earth owes its fertility to water concealed in the air. The comic poet, in consequence, chooses to consider the clouds as the deities of Socrates and seems to have given a general idea of the philosopher's opinion concerning them in one of the choruses in which they are personified.

I shall read an imitation which I hope will at least express the principal sense, if not the beauty, of the original.

> Formed by the potent solar ray
> On the blue bosom of the sea,
> We, while the morning light is hale,
> Glide along the watered vale,
> Or in the midday splendour rest
> Upon the mountain's frozen breast.
> From us, the children of the main,
> Descend the snow, the mist, the rain;
> And our sweet influence shed in dew,
> Gives to the Earth its verdant hue
> And nurtures in the summer shower
> Each tender plant, each fragrant flower.

The representation of the comedy in the theatre of Athens appears to have been the first circumstance which led to the unpopularity and persecution of Socrates. He was present when it was first performed, and he is said to have stood during the whole time that the public might see the object of ridicule of the play. And his answer, when he was asked why he thus exposed himself, equally exhibits the dignity of his mind and the mildness of his character. "I consider myself," says the philosopher, "only as the host at a great feast, and I am happy to be able to provide entertainment for so many people."

Plato appeared to have attended to the observations of
nature less than any of his contemporaries, and in his opinions
concerning the universe he adopted the ideas of the Pythagoreans.
Timaeus seems to have been his great authority, and in his dia-
logue called by the name of that philosopher, he professes to de-
liver his opinions. But he has thrown new clouds of abstracted
metaphysics over doctrines originally obscure, and in examining
the work, the mind is continually perplexed with fancies of numbers passing into quantities, ideas becoming matter, and the
dreams of the [philosopher] forming the universe after the model
of his own intellectual world and fashioning the mind of man
after the image of the material world.

The eloquence of Plato, even in his abstracted systems of
the universe, is of a high character. It may delude and seduce,
but when the beauty of language is taken away, the charm vanishes.
He promises a reality, he presents a dream. His theories, like
brilliant clouds of the evening, may delight and affect the
imagination, but they present nothing permanent. They cannot be
touched, they cannot be retained, and they vanish without leaving in the mind a trace of their existence.

Aristotle, though the disciple of Plato, has paid very
little attention to his opinions concerning nature. Gifted with
a cooler imagination and a more comprehensive and more material
understanding, he condescended to reason on things and to endeavour to explain causes from their effects.

The philosopher of Stagira considered the universe as composed by form added to matter and as capable of being resolved
into air, earth, water, and fire. He made the distinction between simple and compound bodies and supported two principles
of motion, gravity and a repulsive force. This is the most remarkable part of his doctrine and the only one in which he appears to have anticipated the moderns.

Aristotle considered matter as susceptible of any form, and
hence he supposed that different bodies were convertible into
each other by the superinduction of new qualities. It is on
this doctrine of the philosopher that the reasonings of the
schoolmen concerning actual and potential existence are founded.
Aristotle, in advancing it, laid the foundation of those endless
quibbles and sophisms which so much perplexed the minds of men
in the Middle Ages.

And it is on his logic that Thomas Aquinas has founded some
of his ridiculous questions such as whether a potentially existing gold is more valuable than actually existing lead, and
whether a possibly existing angel is better than an actually
existing fly.

It is said by Pliny that Alexander the Great employed large
sums of money and a number of men in procuring subjects in natural history for Aristotle. But amongst these subjects, none of
the fossil productions of the earth are mentioned.[5] Aristotle,
in his views of the globe, seems to have confined himself wholly
either to abstracted speculations concerning the elements, or to
descriptions of the forms and properties of animated being.

Theophrastus, the successor of Aristotle in the Athenian
school, is the first of the ancients who has any claim to the
title of geologist or mineralogist. Theophrastus was a native
of Lesbos, but he left his country at a very early period and
became a disciple of Plato. He was no less celebrated for his
industry, learning, and eloquence than for his generosity and
public spirit.
    This philosopher was the author of a great many important
treatises. He wrote on the earth, the metals, and on stones.
The last only of those three works has come down to modern
times. Its value is great and singular, and it is impossible
not to regret the loss of the other disquisitions upon analogous
subjects.
    In the work on fossils Theophrastus describes with con-
siderable perspicuity a great number of the stones and gems
known to modern mineralogists, and his classification is often
very correct. Thus he divides fossils into three: the incombustible and infusible, the combustible, and the fusible. Under
the head of incombustible and infusible fossils, he describes the
gems and rock crystals. In treating of combustible fossils he
considers pyrites, amber, and different species of mineral coals,
and he speaks of those last substances as being in use. Thus in
the twenty-eighth section of his work he says,

> Those fossil substances called coal that are broken for
> use are earthy; they kindle, however, and burn like wood
> coals. They are found in Liguria, where there is like-
> wise amber, and in Elis, and on the way to Olympia in
> the mountains, and they are employed by the smiths.

He notices several fossil stones and, amongst them, different species of pumice which he considers as formed by volcanic
fires. He was well acquainted with a variety of earths, with
different species of marble, and with the ore of quicksilver, of
lime, and of magnesia. And he mentions a fossil found in the
mines of [Scaptesyle] which inflamed when oil was poured upon it,
a property which has been discovered in no other fossil except in
the black wood of Derbyshire so that there is every reason to
believe that it was an ore of the same kind.
    The geological theories of Theophrastus are often very in-
genious and always very candid. He makes an observation which it
is to be wished were more attended to in modern times, "That the
same effects may be produced in nature by very different causes."
He considered primitive earths of various kinds as the bases of
stones, and he supposed crystals to be formed by solutions of
those earths and a consequent precipitation. He believed in a
peculiar force or attractive power by which the parts cohered together, and he considered the mode of their union, no less than
the peculiarity of the matter of which they are formed, as essential to the differences of the properties of the resulting compounds.
    Theophrastus evidently possessed genius, talent, and the

faculty of minute observation. His influence as a teacher was
great. He is said to have had at one time more than two
thousand scholars, and yet not a single disciple was found to
follow the steps of the illustrious master in the study of
nature, to trust like him to facts, and to record only the results of experience and accurate examination.

The men who called themselves philosophers in the last
period of the declining splendour of Greece paid still less
attention to the sensible properties of things than their pre-
decessors. Epicurus, whose system has made so great a noise,
seems to have borrowed his opinions concerning nature wholly
from Anaxagoras, Leucippus, and Democritus, and even to have
neglected some of the later truths. And the only novelty in
the physical doctrines of the Stoics seems to be their opinion
concerning the successive destruction and renovation of the
earth by water and fire. Zeno is made by Laërtius to maintain
that the great changes in nature by these two powerful agents
takes place as regularly as summer and winter, and that the
earth is destroyed and renovated by uniform laws which operate
in the succession of ages.

> The time will come, says he, when the world will be des-
> troyed that it may be again renovated, when the powers
> of nature will be opposed to each other, when stars will
> rush upon stars and the whole material world now wonder-
> ful in beauty and in harmony will consume away in
> flames.

The Romans, the vanquishers of the Greeks and their decided
superiors in arms, yielded to them the empire of the mind. And
the conquerors were instructed by the conquered and adopted almost all their opinions in literature and in science. The
philosophy of Rome was in fact no more than an imperfect copy of
that of Greece in which some of the original ideas were preserved
but no new ones introduced.

It was at Athens that Lucretius gained all his knowledge
under Zeno and Phaedrus, and he appears to have employed the
whole force of his mind and all the fire of his genius in com-
bining the opinions of the different Greek philosophers and in
contrasting them with those which he conceived to form the true
doctrine as delivered by Epicurus. Even his poetic imagination,
though continually active, seems to have produced nothing new on
the subject of nature, and he attributes all his information to
Epicurus with the most grateful and reverential feelings. In the
beginning of the fifth book he says,

> What verse can soar on so sublime a wing
> As reaches his deserts, what muse can sing
> As he requires, what poet now can raise
> As stately monument of lasting praise
> Due to his merits! his who first explained
> The eternal laws of nature?

Cicero, who revived the poem of Lucretius after his death, studied with great zeal the different systems of philosophy discussed in it, and in his work *On the Nature of the Gods* he has examined most of the ancient theories of cosmogony, but without adding any new opinions of his own. He seems aware of the impracticability of arriving at truth on a subject in which no accurate data were known, and his philosophical disquisitions appear rather to have been intended as amusements of fancy than as serious exercises of reason.

The era of Augustus has been justly celebrated on account of the protection afforded at that time to genius, and on account of the constellation of men of letters that it produced. But yet even in his happy period, Rome could not boast of one man devoted to science, and the philosophy of the age is developed only in the writings of the poets.

Virgil, as would appear from several passages in his work, was a Pythagorean. Horace seems to have favoured the Epicurean doctrine, and Ovid in his *Metamorphoses* has mixed many of the philosophical opinions of the Greeks with their early fables. The only Roman who really deserves the title of an investigator into nature is the elder Pliny. This illustrious person possessed the highest degree of industry and an ardour in the pursuit of knowledge which no difficulties could repress. He considered all the productions of the earth as worthy of attention, either for their order, their beauty, their uses, or relations to man.

Possessed of such requisites for discovery, he was still deficient in the great characteristics of a strong mind and a philosophical spirit. Endowed with a simple heart and apparently incapable of deceiving, he believed almost whatever was related to him; doubt seemed to be a stranger to his understanding. He beheld things in their obvious form with delight and with wonder, and satisfied with what he saw, seldom attempted to refer effects to their causes. Endowed with none of the high elements of reason, with none of those restless workings of the imagination which produced new combinations of ideas, new truths, and new inventions, he was nevertheless a minute observer and a faithful historian, but neither an experimental philospher nor a man of genius.

A small part only of the extensive *Natural History* of Pliny is devoted to geological considerations and to the fossil productions of the earth. The descriptive part of his book on stones is partly borrowed from Theophrastus. But he has added to the simple and philosophical details of the Greek author an infinite variety of unimportant anecdotes, of vague incidents, and a number of ridiculous accounts of the medical virtues of the substances treated. Thus in describing the diamond, after noticing the form of its crystals and its extreme hardness, he details an absurd opinion of its being easily broken by a blow after having been infused for some time in goat's blood. He states there is a great enmity between the diamond and the loadstone, and that this gem is a remedy for madness and an

antidote against poison. And he advertises all ladies, who have
it in their power, to wear diamonds as they are a certain pre-
ventative against melancholy and lowness of spirits.

Pliny gives a minute account of the oriental gems and
states that they are so hard as not to be scratched by agate
and much heavier than crystal. He appears a firm believer in
their medical uses, though he sometimes seems to doubt whether
or not they can be applied to the purposes of enchantment.

There is a strong analogy between Pliny's history of the
precious stones and that given in the Greek poem on gems ascribed
to Orpheus, which, however, is generally believed to have been
written during the age of Constantine. The Greek poet, like the
Roman philosopher, supposes the greatest number of maladies to
be owing to a softness and deficiency of strength in the organic
matter of the body and therefore concludes that the hardest substances in nature are those most calculated to restore it to
health and strength.

Of the metals, Pliny has described gold, silver, copper,
lead, tin, iron, quicksilver, and antimony. And he was acquainted with the ores of zinc, bismuth, and manganese. He has given,
however, no distinct description of the working of the metals,
and he seems to have been wholly unacquainted with the position
of metallic veins and the strata in which they are found.

The most accurate, though perhaps the least entertaining,
part of his book is that in which he describes the great changes
of the globe. He was perfectly acquainted with the origin of
rivers, and one of his most beautiful and finished descriptions
is that of the agency of water, which I have translated from
the thirtieth book.

> What, says he, can more demonstrate the grandeur and
> majesty of nature than the power and the force of water.
> For this one principle, as it were, rules and commands
> all the other elements. It acts upon and dissolves the
> solid earth; it weakens and even extinguishes the devour-
> ing energy of fire; it rises into the air in vapor and
> forms clouds and mists whence lightning and thunder,
> whence hail and snow, and rain, and the fertilizing dew.

Pliny attributes the tides to their real cause, the effects
of the sun and moon. Earthquakes, he considers as owing to elastic fluids disengaged in the interior caverns of the globe, and
inundations, as depending in many cases upon the sudden elevations
of land in the sea.

Pliny flourished in the time of Vespasian and Titus. He
lived honoured by those emperors and respected by the public.
He is one of the few philosophers who enjoyed riches and reputation in his lifetime, and his death may be said to have been
glorious and worthy of his character. He was commander of the
Roman fleet at Misenum. In the seventy-ninth year of the Christian era, an eruption of Vesuvius took place, the first and perhaps the greatest. Animated with a desire of examining this won-

derful phenomenon, he landed from a small vessel on the coast
and, whilst engaged in making his observation, was surprised by
the streams of ignited matter flowing rapidly from the mountain.
He tried to escape, but all retreat was impossible and he fell
suffocated in an instant by the sulphurous vapour emitted from
the lava.

With this account of Pliny I shall conclude the present lecture in which I have endeavoured to collect from a variety of
sources the little knowledge that the ancients possessed concerning the earth, its elements, and its changes.

To bring forth truth from the darkness of antiquity is a
much more delightful task than to detail errors and absurdities,
but the knowledge of the present day will perhaps appear with
greater brightness and distinctness in being contrasted with
the ignorance of elder periods. In literature, in the fine
arts, in works of taste, and in all combinations that required
only the native powers of genius, the ancients perhaps are still
our masters. But in philosophy, their systems and their doctrines relating to nature, when compared with the truth and
theories of modern time, appear as the vain toys and amusements
of children when contrasted with the useful occupations and pursuits of men.

Lecture Three

The science of nature depends in a great measure upon artificial inventions for assisting the faculties of the mind and for increasing the activity of genius. And it requires for its advancement a state of society in which active patronage is offered, and in which it is assisted no less by the diffusion of letters than by the application of industry, riches, and power to its numerous and diversified objects.

It is not therefore extraordinary that the spirit of observation and of experiment, which maintained with difficulty a sickly existence even in the most splendid and happy eras of Rome, should have disappeared altogether at the time of her decline and degradation, when turbulence at home, feebleness abroad, and indulgence in the grossest luxury had prepared the way for the total downfall of the state.

Egypt, recorded as the parent of science in the most remote antiquity, was destined by a singular concurrence of circumstances to become her nurse in a later period. And philosophy, driven from the West by persecution, by the ravages of barbarians, and the tumult of savage warfare, found a transient refuge within the walls of Alexandria. Towards the close of the fourth century, at the time when the Goths were conquering the finest provinces of Europe and keeping the Romans in continual alarm, this city enjoyed a profound peace and afforded an asylum to students from all parts of the Empire.

The school of Alexandria produced several laborious geometricians, but the greatest ornament belonging to it and the most illustrious philosopher of the age was Hypatia, daughter of Theon. This celebrated lady is said to have been equally

distinguished for her skill in mathematical and in general
science, and in the knowledge of nature and of the earth. And
that she applied herself to experiment is evident from one of
her inventions, the hydrometer, the instrument now in common
use for ascertaining the relative weights of fluids.

Hypatia taught after her father's death in her native city.
By the eloquence and soundness of her instructions she excited
the highest zeal for moral and intellectual improvement amongst
her disciplines. She was no less admirable for the grace and
loveliness of her person than for talents and virtue, and she
shone with greater lustre being, as it were, a single brilliant
star in a night of clouds and obscurity.

Learning flourished at Alexandria for nearly two hundred
years, but literature and metaphysics seem to have been more
attended to than experimental knowledge. Of the improvements
made, however, we are almost wholly ignorant, for the same cause
which put a stop to the future progress of science destroyed
all the records of the advancement that had been made. The
city was conquered by the Saracens, and the most magnificent and
perfect library in the world was burnt at the command of Omar.

Philosophy and letters were taken away from the people, and
they were obliged by the force of sword and of fire to substi-
tute for them the absurdities and impostures of the Koran. All
was again ignorance and darkness till the eighth century when a
feeble gleam of light burst forth in Arabia, and the patronage
of two celebrated caliphs, Al Mamun and Al Mansur, kept alive
for a short time the spark of enquiry, but at this period there
were no permanent materials for it to act upon. The spirit of
science and that of Mahometanism, like the good and evil principles of Zoroaster, were incapable of being reconciled. Improvement languished in the reign of the later sovereigns of Baghdad
and it soon perished altogether with the Empire, which was one
of the first that fell beneath the rising power of the Turks.

Very few works of importance have been transmitted to us
from the Saracens, and details that we are in possession of with
regard to the short literary history of this people are principally valuable on account of their connection, about the time
of the Crusades, with the establishment of a particular philosophical sect in Europe which had the greatest influence upon
the progress of natural science, of chemistry, and of geology.

Geber and Rhazes of Baghdad are said to have been the first
men who pursued the operations of chemistry with the view of
ascertaining the natures of the bodies belonging to the earth
and for effecting changes in them. And they are the philosophers to whom the earliest alchemists of the West ascribe the
origin of their art.

Alchemy as at first professed was considered as a universal
science. It was regarded not merely as affording the means of
converting other metals into gold and as providing remedies for
all maladies, but as laying open the great secrets of nature and
as explaining the phenomena of the earth, of the sea, of the
air, and of the subterranean mineral kingdoms.

Albertus Magnus and Roger Bacon, who are the first persons
recorded as applying themselves to experiment in the Dark Ages,
seem both to have been guided by very philosophical views and to
have had as their principal objects the advancement of science.
The elder Bacon confesses in his great work that he gained the
foundation of his knowledge from the Arabian writers upon whom
he bestows high and unqualified praise. This great man evidently
studied nature and the productions of the earth with the views of
a philosopher, but his knowledge was so superior as to be unintelligible in the age in which he lived. The wonders produced by
chemistry were referred by the people to the agency of evil
spirits, and a very short time after he had written a book to
prove the nonexistence of magic, he was himself persecuted as
an enchanter. He was imprisoned in 1278 at the command of the
Principal of the Franciscans for having brought the Order into
disrepute by pretending to natural wisdom and by exercising unholy and supernatural powers.

Roger Bacon appears to have made use of the philosophical
and experimental method of the Saracens without following any
of the absurdities of their doctrines. He had seen admirable
changes produced in bodies and he knew not the limits of the
operation of nature. By him the production of gold and the
transmutation of metals were considered only as objects worthy
of investigation, and he never affirmed confidently concerning
them. He had made many important discoveries and particularly
that of an explosive compound similar to gunpowder; yet he seldom suffered his imagination to cloud his reason, and he was no
enthusiast with regard to the merit of his inventions.

He possessed the modest and dignified feeling of science,
but most of the alchemists who flourished in the next century
were of a very different complexion. They formed themselves
into a fraternity. They professed to be in possession of great
and important secrets. They connected a peculiar mysticism with
their philosophical doctrines. They attached to alchemy a language similar to that which had been employed in the Platonic
philosophy, and they pretended not only to a knowledge of the
materials of the globe and the changes of things, but likewise to
an acquaintance with the elements and the spiritual powers by
which they supposed these were governed.

The Arabians had blended in their philosophy many brilliant
superstitions concerning the agency of a class of beings supposed
to be superior to man and possessed of great knowledge and of
high power. The notions of fairies or genii which have been de-
picted with so much vividness of fancy and liveliness of description in the *Thousand and One Nights* seems to have been adopted
under a new form by the adepts of this period in Europe who,
though they altered the denominations of the spiritual agents
created by the fancy of the orientalists, still attribute to them
similar attributes and similar relations to man.

Agrippa [1486-1535; DSB 1:79], in his work on the occult
sciences, endeavours to develop a general system of the philosophy
of the earth, and he begins by supposing that all things on the

terrestrial globe are under the protection of influences residing in the stars and of good and evil genii, and that true knowledge can be derived only from a communication with those sprit-
ual beings.

Agrippa considered fire as the great active principle in
nature and from its operation upon earth, which he regarded as a
passive element, he conceived the great appearances of the external world to be produced. Water and air he thought were
not capable of being converted into each other, and yet he considered air as a vital principle which existed in almost all
substances and as essential to their composition and their nature.

This last passage is one of the few philosophical ones
in his work most worthy of notice for his pages are generally
filled with the grossest absurdities. It is difficult to conjecture from them whether he is self-deluded or whether he is
endeavouring to deceive others, whether his imagination was
distempered or whether he was attempting an imposture.

Agrippa was born at Cologne in 1463.[1] After the publication of his system, he was of course considered as a magician by
the vulgar and a number of idle tales exist with regard to his
history. He asserted that he was in possession of the philosophers' stone, yet he was always miserably poor. He affirmed
that he had spirits under his command who were capable of con-
veying him from place to place, and yet he passed the last years
of his life in the prison of Lyons where he was confined not because he professed magic, but because he had violently abused
the Queen Mother of France.

The general systems of the famous Paracelsus [1493?-1541;
DSB 10:304][2] seem to have differed very little from that of
Agrippa. He supposed the elements under the influence of the
stars and that all the spaces in the universe void of matter
were filled with peculiar spiritual beings. And he has made
use of the very names which have been so much employed in the
machinery of modern poetry. Thus he regarded the sylphs and
gnomes as belonging to the earth, spectres to the air, and sala-
manders to fire. Paracelsus introduced sulphur and mercury into
his theory as principles composing the metals, and he regarded
animals as chiefly constituted of air, and vegetables of earth.

The life of this man, who was born in Switzerland in 1493,
was full of extraordinary events. He travelled through most of
the countries of Europe and Asia and even passed a considerable
time in Africa, seeking, wherever he went, for secrets in chemistry and professing to cure all maladies. His enthusiasm almost supplied his want of genius, and confidence in his own
powers was [probably] more impressive, in [his] age as well as
in the present, than common sense.[3]

He was appointed to the professorship at Basel but he soon
contrived to disgust the magistrates and to gain dismissal. He
was fond of living only amongst the lower order of the people,
and from associating continually with his inferiors he acquired
the most offensive vanity, the most insupportable arrogance. He

abused the physicians of his time in the grossest terms. He
substituted chemical preparations for vegetable medicines. And
he publicly burnt the works of Avicenna saying that there were
no other masters in medicine but nature and himself.

He passed his last days in wandering about Germany, and
when he was reproached on account of his way of life he proudly
answered it was not fit for a person born to succour the whole
of mankind to attach himself to any one spot, and he who was
destined to read the book of nature ought not to have his eyes
always fixed upon the same leaf.

Van Helmont [1579-1644; DSB 6:253][4] of Brussels, like
Paracelsus, was a mystic, but he combined much more sound philosophy in his views of nature and of the earth. He was a dili-
gent experimenter and a minute observer, and some of his researches led to important results. He was the first person who
supposed that elastic fluid existed different from the air of
the atmosphere and from him we derive the modern chemical name
"gas." He likewise invented another denomination, "blas," which
is not quite so intelligible. He considered it as of a much
finer nature than air, and as being derived by the earth from
the sun and stars. Van Helmont made some very laborious experiments to prove that water was capable of being converted
into earth, and he revived the doctrine of Thales that it was
the general productive principle.

The sixteenth century was the great era in which the mystical adepts flourished. The famous Rosicrucian Society seems to
have been established about this time by them, and the history
of this institution is fully as obscure and dark as that of the
kindred sect of the modern Illuminés.

It would be of little use and foreign to the objects of
these lectures to enter upon an examination of all the writings
published on alchemy in different parts of Europe by Valentine,
by Fludd, by Böhme,[5] and others, and I should not have given any
sketch even of the doctrines of the adepts had not some of their
labours been intimately connected with the progress of true
chemical and geological knowledge.

They were experimenters with false views, but they nevertheless sometimes developed truth. They were indefatigable in
their researches and they had ascertained a number of the most
curious properties of bodies without being able to account phil-
osophically for their effects. Though visionaries, they sometimes acted according to the dictates of right reason, as persons who walk in their sleep sometimes perform in a proper manner the common occupations of life. The facts scattered amongst
their absurdities were valuable in the highest degree to the
philosophical men who came after them, and a true and enlightened science arose by degrees and amidst clouds and obscurity
from the furnaces of the alchemists.

George Agricola [1494-1555; DSB 1:77] was the first person
who published in practical mineralogy and in geological chemistry in the sixteenth century.[6] His work on the metals contains
a number of important truths which are always detailed with pe-

culiar candour and modesty. He describes the general position
of the veins in the great mineral countries of Germany. He
gives a natural history of ores and a particular account of the
manner in which they are worked, and in none of his details
does he enter upon wild speculations. And he has treated the
professed adepts of his age with a cool and dignified indif-
ference.

Lazarus Ercker [1530?-1594; DSB 4:393], assay master general to the German mines, was contemporary with Agricola. His
work on the art of assaying has been translated into English by
Sir John Pettus and it contains much valuable matter with regard to metallurgical chemistry.[7] Both Agricola and Ercker considered the metals as incapable of being transformed into each
other and they regarded them as produced by certain operations
of nature which cannot be imitated by artificial means.

The period in which these two mineralogists wrote was the
great era of improvement in the general methods of the natural
sciences. It was the period in which the death blow was given
to alchemy by Gassendi and Kepler who both answered Dr. Fludd,
the last public defender of the Rosicrucians.[8] It was the period
in which Galileo pursued the paths of experimental enquiry, and
in which the second illustrious Bacon had the honour of establishing the sciences upon their true and immutable foundations.

Many scientific persons before Bacon had pursued the method
of experiment in all its precision; many had dared to despise the
logic and forms of the ancients. But he was the first philosopher who laid down plans for extending knowledge of universal
application, who ventured to assert that all the sciences could
be nothing more than expressions of facts and that the first
step towards the attainment of real discovery was the humiliating confession of ignorance.

Bacon was prepared by nature, by education, and by his
habits of study for effecting the great revolution in philosophy.
His knowledge was extensive, his instances were copious. His
genius was equally capable of developing the lighter and the
more profound relations of things. He possessed a strong feeling, but it was uniformly directed by reason. He was gifted with
a vivid imagination, but it was modified by a most correct taste.

The influence of rank and of situation assisted his views.
The public was prepared to receive them, and he was enabled to
develop his opinions with full confidence that they would be
adopted with reverence, even in his own time, and that they would
carry his memory into future ages with great and unchanging glory.

The pursuit of the new methods of investigation, in a very
short time, wholly altered the face of every department of natural knowledge, but their influence was in no case more distinct
than in the advancement of geology and chemistry. Though much
labour had been bestowed upon these extensive fields of investigation, they had hitherto, as we have seen, been little productive. Speculation had been misplaced, observation confined,
and experiment principally directed rather towards impossible
than to practical things.

In the novel system, hypothesis was excluded except as a guide to actual trials. Combinations of thoughts were considered as truths only when conformable to nature and not when they merely expressed the caprices of the imagination. And those enquiries only were regarded as valuable which were made upon the hidden sensible properties of things and upon the existing relations of facts.

One of the first persons who applied himself with zeal to the investigation of the theory of the earth with enlightened views and distinct plans was Robert Hooke [1635-1702; DSB 6:481]. This illustrious man distinguished himself at a very early age by his mechanical knowledge, and in his chemical speculations he developed, in a most consistent and penetrating view, the general theory of combustion. His genius was too powerful and too active to be limited to these considerable objects, and in the course of his scientific career he applied his great abilities to almost all the principal objects of investigation in natural science.

He devoted his attention equally to the great laws of the globe considered as a part of the planetary system, and to the minute changes and imperceptible alterations of the surface. His opinions upon terrestrial motions are marked by a singular [omitted word(s)]. And his hypothesis of earthquakes and volcanoes is scarcely exceeded by any of the refinements of theory in modern times. Few men have done more than Hooke and few have been more neglected. One of the principal reasons is that he was the contemporary of Newton whose just glory and whose spendid discoveries concerning the great laws of the universe absorbed the attention of the philosophers of the age and made them almost regardless of inferior inventions.

Hooke, in the geological speculations which were published after his death,[9] assuming as a fact the existence of shells and the remains of marine animals in rocks and in hills of considerable elevation, draws the conclusion that these strata must have been at some period covered by the sea. And after reasoning upon the general effects of earthquakes and of volcanic explosions, he infers that there are no causes in nature more probable than these for the elevation of land, of rocks, and of mountains from the bosom of the waters.

He brings forward many chemical instances to show that water is capable of dissolving and of depositing stony matter and that its agency is fully adequate to the formation of strata. And he shows that both in modern and in ancient times several cases have occurred of islands and of tracts of land rising out of the sea in consequence of submarine volcanoes. The ideas of Hooke are conceived with boldness and developed with precision. In many principal parts of theory he had anticipated the modern plutonic philosophers. He seems, however, to have restrained his imagination. He has contented himself with demonstrating a certain part of the existing order of nature without attempting a general system of the past changes of the globe and of its future destiny.

Becher [1635-1682; DSB 1:548] lived at the same period as Hooke and, like Hooke, he devoted his life to science. But he selected one department and directed to geological chemistry the whole of the powers of his mind.[10] In his book called *Subterranean Philosophy*, published in Latin in 1669, he has given some very beautiful details with regard to the mineral kingdom and has offered views of a theory perfectly new. His ideas concerning combination and decomposition are of the most accurate nature; his notion of an inflammable principle formed the foundation of the hypothesis of Stahl [1660-1734; DSB 12:599].[11] He gave to chemical philosophy new dimensions and new instruments of investigation, and he attempted to form a general theory in which all the operations of nature should be explained and illustrated by experiments of human invention. Becher, with the sound judgement of a true philosopher, assumes a certain primary order of things resulting from the creation of the globe, and he applies his reasonings only to the solution of the phenomena of the changes which have since taken place.

Water in the mineral kingdon and elastic vapours, he considered as the great agents which have altered the position of strata and the direction of metallic veins, and from the decomposition of certain rocks by these natural agents he supposes that new ones have been formed of a much softer nature and fitted to become soils and to support vegetable life.

Becher was born at Speyer in 1645 [sic]. He was for some time a professor of medicine in Bavaria, but he was driven from Germany by envy and persecution. England became his adopted country. He was patronized by the great Boyle, in consequence of whose liberality he was enabled to study mineralogy in Cornwall, to pursue his experimental researches, and to produce some of those works in natural science which will always confer honour on his name.

Hooke and Becher had contented themselves with partial views, and the celebrated Leibniz [1646-1716; DSB 8:149] is the first philosopher who attempted a general system of geology. In his *Protogaea* published in 1693[12] he enters upon the great doctrines relating to the globe, the primary creation, the Deluge, the changes that have been since produced and those that will be produced in future ages. He assumes as the foundation of his reasoning the account of the creation given in the sacred writings.

He attempts to show that the globe at its origin must have existed in a state of igneous fusion, in consequence of which it assumed its spherical form. The heavier substances such as the metals, he supposed to have occupied the centre; and he conceives that the rocks which crystallized at first in regular strata were afterwards broken and raised into mountains by explosions taking place from beneath, and the chasms produced and filled with metallic veins. The water, which at first existed as highly rarified vapour, he supposes to have been by degrees condensed; and from its action in its heated state upon the land he conceived that a part of the surface was decomposed, broken down, and converted into a soil on which, by the Almighty fiat, vegetables and animals were made to exist, and the whole clothed with beauty and rendered subservient to the purposes of the last created and most perfect of beings.

Leibniz supposes that there may have been two causes for the Deluge: the elevation of land in the former bed of the ocean in consequence of explosions from the central ignited mass, and the passage of the tail of a comet through the atmosphere, loaded with aqueous vapour. By the first cause he explains the passage in Genesis, "The fountains of the great deep were broken up;" and by the second, [he explains] the other sentence [also in Genesis], "the windows of heaven were opened and the rain was upon the earth for forty days and forty nights."

Leibniz supposes that the principal changes which have taken place since the Deluge have depended either upon volcanic fires of great inundations from the ocean, and he conceives that fire and water are counterbalancing powers by which an order of things similar to the present will be maintained till the great and general conflagration. Such is the doctrine of the *Protogaea* as found in the excellent edition of the works of Leibniz by Mr. Dutens.[13] The work itself is in the highest degree deserving of a minute examination. It is worthy of the genius of the illustrious author. When facts are stated in it they are detailed with clearness and accuracy, and when hypotheses are indulged in they mark an imagination fertile in novelty and pregnant with bold conceptions.

That his reasoning is not always convincing rather depends upon his endeavouring to solve problems concerning which there are no data, than upon any imperfection in his methods. The subject which he has undertaken to discuss is one almost above the powers of the human mind. It required high talents to produce any philosophical views with regard to it and the abilities of Leibniz, where they have not developed actual truth, have uniformly produced some brilliant semblance to reality in which the fancy always and sometimes even the reason can dwell with delight.

The *Protogaea* has been very little known but many writers appear to have borrowed from it. Whiston [1667-1752; DSB 14: 295] adopted the idea of a comet having produced the Deluge and has endeavoured to prove it from mathematical deductions.[14] His treatise is, however, very much inferior to precision to that of the Prussian philosopher and contains no ideas remarkable either for strength or novelty.

Another system which appeared about the same time as that of Leibniz has excited infinitely more attention in Britain; I mean the *Sacred Theory of the Earth* by Dr. Burnet [1635?-1715; DSB 2:612].[15] In his work the globe is supposed to have been created without irregularities, perfectly spherical and smooth, having neither hills, mountains, nor valleys. The land is supposed to have been equally diffused above the ocean and to have covered it as a thin crust, and the Deluge is considered as

having been occasioned by the breaking down of the land into the
great abyss beneath, an event which he supposes to have produced all the varieties of sea and earth that at present exist,
the formation of islands, and the great elevation of continents.

Burnet paints with all the feeling of an enthusiast the
delightful state of the inhabitants of the globe before the Deluge. In his hypothesis, as there could have been no variety of
seasons, he conceives that they must have enjoyed a perpetual
spring, that the face of heaven, and the sun, and the stars must
have been exhibited to them in unclouded beauty, and that the
whole of the earth must have been green with perpetual vegetation.

Burnet saw none of the errors of his system. He did not
perceive that without the action of the waters of the sea upon
the atmosphere there could be no rain, no dews, no rivers, and
consequently, no plants and no animal life. His mode of accounting for the mountains of the earth by the filtration of water
through the solid matter and its elevation into the air is perfectly inadmissible. And everyone must allow that the feelings
and taste of the antediluvians must have been very different
from ours if they could derive more pleasure from the contemplation of a flat surface and a uniform sky than from the view of
the variety of hill and dale, of mountain and river scenery, and
the ever varying appearances of clouds.

Dr. Keil [1671-1721; DSB 7:275], in his examination of the
Sacred Theory,[16] has distinctly proved that it is wholly incompatible with the general laws of the globe and not conformable
in the slightest degree to facts. As a poetical romance it is
highly interesting and entertaining, but as a philosophical
opinion it is wholly unworthy of attention and it affords one
proof amongst many others of the folly of attempts to wrest the
meaning of the sacred writings, which express general truths,
so as to make them serve as supports for hypotheses of human invention, so as to blend them with the visions and fancies of men.

The last system which I shall mention in this lecture is
that of Buffon [1707-1788; DSB 2:576], one of the most singular
and extraordinary productions of the human mind.[17] Beauties and
defects are blended in it and it is implete with ingenuity and
with errors.

Buffon's idea of the primary state of the globe is wholly
borrowed from Leibniz. He conceives it to have existed in
igneous fusion, but despising the modest opinion of the author
of the *Protogaea* that such was its primary creation, he supposes
it to have been produced from the ruins of another system, to
have been carried off from the sun by the powerful impulse of
a comet, and to have been moved into its present orbit by projection and the force of gravity.

He fixes the era at which this great event is supposed to
have taken place at a distance of time quite inconceivable. And
he imagines three different epochs in the history of the cooling
of the surface before marine animals existed and five epochs before it was fitted for the residence of land animals and of men.

Buffon brings forward two principal arguments in support of his hypothesis, and they are now both known to be founded upon false principles. The first is that the temperature of the earth is found higher the farther we penetrate beneath the surface. The second is that the surface of the earth is composed of matters which are almost all capable of being vitrified by heat.

Now it is well ascertained that, except in volcanic countries, the temperature of the earth at a considerable depth is the mean temperature of the atmosphere, and that it varies according to that mean in all the different climates, so that it is evidently produced not by internal but by external causes, not by a central fire but by the solar rays and the heat of the air.

With regard to the other propositions, it is certain that the greater number of our rocks are wholly incapable of being fused by any heat that we are able to apply to them. And though it is not improbable in a temperature such as that of the sun they would become liquid, yet we have no right to make inference merely for the sake of forming a theory. For a theory ought in all cases to be not an expression of conjecture but an arrangement of facts.[18]

The ideas of Buffon concerning the changes of the globe that are now taking place and the general order of nature are scarcely less bold or less visionary than his notions concerning the origin of things. He conceives that the land and the sea are continually changing places, that from the action of rains, of rivers, and of torrents, the solid matter of the land is constantly divided, decomposed, and deposited in the ocean, in which, carried by currents towards the equator, it is again accumulated in strata and formed by the powers of aqueous consolidation into new rocks and new mountains which rise at length above the waves and offer a due compensation for the destruction of parts of ancient continents and islands.

This hypothesis, in which the waters of the atmosphere are supposed to be the destroying cause and the waters of the sea the regenerating cause of the solid land, is supported by the author with considerable address and ingenuity. He endeavours to prove from the shells and remains of fishes found in some parts of high mountains that the present continents must have been formed in the bed of the ocean. He endeavours to show that some of the islands under the equator have been very lately raised. And he brings forward many accurate instances which demonstrate that the sea has gained upon the land to some extent in many different parts of Europe.

I shall not at present enter upon a full discussion of these opinions, for they will again occupy our attention in the examination of a much more comprehensive and more artfully constructed theory, that of Dr. Hutton. I shall merely offer a few general observations upon them.

With regard to the idea that our present land was formed in the bosom of the sea by the mere action of water, it is one that may be controverted and overturned with ease. The greatest portion of our high rocks and elevated mountains are crystallized masses containing no organic remains and not arranged in layers, and they consist of matter wholly incapable of being consolidated by water, such as granite, which forms a considerable part of the Alps and which constitutes the great mountains of Scotland and the hills and cliffs of Cornwall. And stratified rocks, even which seem themselves to have been deposited by water, are often blended with rocks of another description so as rather to indicate. a great inundation and rapid transit of the ocean over the land than a gradual and quiet deposition of the solid matter of the earth in the bed of the sea.

It is impossible to doubt that in the lapse of ages some slight diminution of our mountains must take place from the agency of rivers, but there are many causes in nature which appear to counteract the effect. Forty centuries nearly have passed away and no material change in the form of the solid matter of the globe has taken place in consequence of this agency.[19] From the most accurate observation it appears nearly a hundred years would be consumed before a single inch of a rock like that which this instance represents, or of Mount Blanc, would be lost. And with regard to the ultimate termination of such effects, it is almost ridiculous to reason seriously.

Man attached to the globe has a limited sphere of action, limited faculties, and but a short period for the employment of them. He was not intended to waste his time in guesses concerning what is to take place in infinite duration, but he was rather born to reason from the past and the present concerning his immediate and future destinies. In philosophy, as well as in common life, he ought only to be guided by certainties, by distinct probabilities, or by strong analogies.

To conclude, after censuring M. Buffon's wilder speculations, it would be unfair to deny him the highest praise for order of arrangement, spendour of description, and brilliancy of style. He has adorned a subject of the most abstracted kind in the refined charms and elegancies of literature, and he has bestowed on it an interest which it had never possessed before. His merit in this respect is transcendent. There is nothing highly difficult in the task of delineating human character, in painting the passions and affections of the mind with eloquence and with feeling, for these belong to all minds, all climes, and all ages, and their effects are universal. But to clothe the dull and dead objects of a speculative branch of science in the forms of beauty and to infuse into them animation and life, this is indeed a work which the strongest genius only can accomplish. And it is a work in which Buffon has perfectly succeeded.

Lecture Four

It has often happened during the progress of science that
two doctrines immediately opposite have been supported with
great powers of reasoning and almost with equal plausibility.
Such a circumstance, however, almost always marks the infancy
of the study and the imperfection of the methods of investigation. It proves that speculation merely has been employed or
partial facts only examined, and the philosophical enquirer, in
opposing supposition to supposition, and error to error, will
instantly perceive the insufficiency of either class of opinions, and will thus preserve the mind free from doubt, from
the prejudices of false systems, from mistakes and confusion.

This method may be sometimes applied with great success
in the examination of geological theories. It removes the
necessity of strict discussion, and it often develops absurdity
without the assistance of laborious logical deductions.

Buffon, in his theory of the globe, supposes that the
surface is constantly becoming colder. M. de Maillet [1656-1738;
DSB 9:26], on the other hand, conceives that it is daily grow-
ing warmer. The first attempts to prove that ice, in the
course of ages, will extend even to the countries under the
tropics, whilst the last conceives that in no very distant
period it will disappear even from the poles.

The one author fears that unless the energies of the sun are
capable of opposing the laws of nature our cultivated lands will
become snow clad plains. And the other dreads their conversion
into burning deserts of sand, incapable of supporting vegetable
life.

It is scarcely possible to imagine two hypotheses in more
perfect opposition; both have been defended with ingenuity, both

are founded upon mistakes. The one may be balanced against the
other; there will be scarcely a perceptible difference in their
weight and importance, and they may safely be given up together.

The work of M. de Maillet on geology is called *Telliamed*,
which is an anagram of his name.[1] A translation of it appeared
in English in 1750. The author professes to develop the opinions of an Indian philosopher whom he had met at Cairo when the
consul general of France was in that city. Though he intended
to produce a philosophical theory, he has given it in the form of
a romance, for the fables are more numerous than the facts, and
it affords much entertainment, but little or no instruction.

M. de Maillet, from having observed an increase of land
on several parts of the coast of the Mediterranean, conceives
that there must be a uniform law of diminution of the sea, in
consequence of which the whole surface must at length become
dry land. His opinion is very similar to that ancient one of
Leucippus. And he further supposes that the whole of the globe
in early ages was covered with water, and that from the slow
and gradual disappearance of the ocean our present islands and
continents have been formed and the existing order of things
produced.

I have mentioned this hypothesis because it is the foundation of the idea of the gradually increasing temperature of the
land, for M. de Maillet justly conceives that, if water disap-
pears, fire must accumulate and that the dry surface must soon
become uninhabitable.

All his observations of facts, however, are partial, and
his notion of the dissipation of the moisture of our globe into
space shows a total ignorance of the laws of evaporation, which
long before the time of the publication of his work had been
fully developed by Dr. Halley [1656?-1743; DSB 6:67].[2]

This excellent philosopher has proved from the most
accurate data that the fluid raised from the ocean and the
surface of the earth in consequence of the solar heat is condensed by cold in the upper regions of the air, and that from
its deposition clouds are produced, and dew and rain, [all of]
which water the earth, support vegetable and animal life, and
form rivers that return the superfluous moisture into the sea,
its parent source. According to this beautiful system, the
foundation of which are distinct truths, there is a constant
circulation of moisture upon our globe; no portion rises above
the atmosphere, no portion is lost; and amidst a variety of combinations and decompositions the equilibrium is preserved, the
quantity is unimpaired, and there is a constant application of
it to the uses of living beings.

Most of the important errors of the early geological systems
of the moderns may indeed be referred to a want of acquaintance
in their authors with the great facts belonging to the order of
nature and to an ignorance of the real essence of the subterraneous kingdom. Chemistry and mineralogy till the middle of
the last century were merely in the embryonic state without
order or beauty. And they offer the only true and permanent

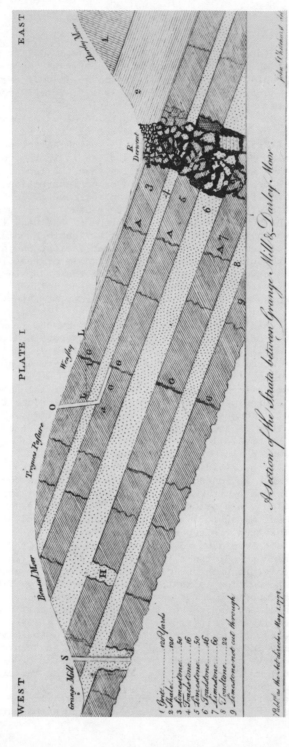

Figure 4.1. A Section of the Strata between Grange Mill and Darley Moore. The toadstone strata (4, 6, and 8) are represented as lying between the limestone strata (3, 5, and 7) and as interrupting the veins in the limestone at G, G, G, and at A, A, A. From John Whitehurst, *Inquiry into the Original State and Formation of the Earth* (London, 1778), pl. I.

foundations on which a just theory of the earth can be erected.
Not even the genius of a Becher or a Hooke, not even the minutely
enquiring spirit of a Woodward [1665-1728; DSB 14:500],[3] nor the
imagination of a Buffon could supply the want of an extended
system of experimental knowledge.

And it was not till chemical analysis had discovered the
real composition of mineral bodies and given accuracy to the
systems of classification, it was not till this period, that
the changes taking place in the surface of the globe in the present time could be accurately ascertained or any true analogical
evidences gained of the intimate nature of the revolutions which
had occurred in past ages.

One of the first persons who endeavoured to employ real
mineralogical and chemical knowledge in the solution of the most
remarkable appearances of the surface of the globe was the late
Mr. Whitehurst [1713-1788; DSB 14:311]. His enquiry into the
state and formation of the earth was published in 1778, about
the period when some of the most brilliant discoveries ever
made with regard to the intimate nature of bodies were brought
to light by Dr. Black [1728-1799; DSB 11:173], Mr. Cavendish
[1731-1810; DSB 3:155], Dr. Priestley [1733-1804; DSB 11:139],
and Mr. Bergman [1735-1784; DSB 2:4].[4]

Whitehurst was evidently gifted with an undefatigable spirit
of investigation similar to that which actuated those celebrated
men. And his labours, though directed toward objects much more
obscure and more difficultly attainable, have been highly meri-
torious and productive of important results.

Whitehurst passed the early part of his life in Derbyshire,
and it was from an observation of the many interesting phenomena
of the mineral kingdom which occur upon a great scale in that
county that he was first induced to direct his attention to geology; not so much, as he himself informs us, with a view of con-
structing a theory as with the hope of obtaining such a competent
knowledge of subterraneous geography as might be applicable to
useful purposes and to discoveries in mining. His attention was,
however, soon arrested by a fact which it was scarcely possible
for an active mind to notice without indulging in speculations,
without reasoning from effects concerning causes.

The principal strata in Derbyshire are sandstone, slate,
coal, and limestone. In some cases they are parallel to the
horizon and regular in their disposition, but in other cases
they are considerably broken and inclined. In some districts
of the county a substance is found called by the miners, toadstone, which is in fact a species of basalt.[5] And where this
rock occurs in the greatest quantity, the stratified masses are
usually found most deranged and confused in their position.

Sketches of the sections of the strata will perhaps illustrate the history, and a single glace of the eye will be more
convincing than the most minute description.

Instances

It was in examining this curious phenomenon that Whitehurst
was led to form the conclusion of the connection of the dis-

order of the strata with the presence of the toadstone. Finding that this rock was lower than the others, and that no metallic veins existed in it, and that even the veins of the limestone
when intersected by it appeared on the opposite sides, he concluded that the strata were of prior formation and that this substance has been interfused amongst them in a liquid state by some
great convulsion of nature, which had produced chasms capable of
receiving it, and which had broken and dislodged the sandstone,
slate, and limestone.

The toadstone of Derbyshire and the lavas of Vesuvius and
Etna are very similar in many of their external properties. This
resemblance did not escape Whitehurst and it afforded him new
ideas and new analogies. Lava is certainly formed by the action
of heat upon rocks concealed in the very bosom of the earth. It
immediately occurred to our author that toadstone must have had
a similar origin, and when he had conceived this, his explanation
appeared complete, for it was easy to conceive that the same volcanic fire which produced the ignited and liquid mass would be
fully adequate to the derangement of the strata, to the separation of veins, and to the formation of caverns in the superincumbent mass. All the facts seemed to combine; all were subservient to one demonstration. Fully satisfied with the truth
of his early speculations, with their apparent conformity to
nature, Whitehurst did not long suffer his imagination to rest
upon the first aspect of things; new objects of discussion
arose, new elucidations of them, and partial theories gradually led to general hypothesis.

Most of the stratified rocks of Derbyshire contain shells
or the remains of marine animals. In considering this circumstance, Whitehurst was led to form the conclusion, almost
common to scientific men, that they had been covered by the
sea, and he immediately connected the cause of the Deluge
with their present elevation, supposing that they had been
raised from the bottom of the ocean by the same fires which
he conceived to have produced their irregularities. During
the rise of new continents and islands assumed to have taken
place, it necessarily follows that the waters must have flowed
in upon the lands already existing. From so great a change
in the surface, the author conceives that an alteration in
the centre of gravity of the earth must have occurred, and
that in consequence at this period its axis may have gained
its present inclination to the plane of its orbit.

Such appears to have been the progress of the mind of
the philosopher of Derby in his investigations. Though in
his work the results are stated in a different order, his
reasonings are evidently all deduced from particular facts.
But in developing his system he commences with the statement
of general principles, and he afterwards applies them to the
phenomena.

It will not be essential in our present enquiry to follow
him minutely through the mere suppositions and extended parts
of his theory in which he reasons concerning the creation and

the original state of the globe. A few words will sufficiently
explain his notions.

Guided by the Newtonian laws of gravitation, he ascertained
that, conceiving the earth to have been created as a fluid or
soft mass, it must have assumed on becoming solid the form which
at present belongs to it, that of an oblate spheroid, flattened
at the poles.

He supposes it to have been covered with water and peopled
with marine animals immediately after the creation. And he conjectures that islands were gradually formed in the great ocean in
consequence of the action of waves and currents, which islands he
considers as the abodes of the antedeluvian race of men.

He supposes like Burnet that before the Deluge the axis of
the earth was perpendicular to the plane of its orbit and consequently that there was no variety of seasons and that days and
nights were equal upon every part of the surface.

In his work a plate is given of the form of the primitive
islands raised at a gentle elevation from the sea.

This design represents the section. [Figure 4.2, overleaf.]

These islands, the fancy of Whitehurst has pictured as
abounding with all the richest productions of nature, as equally
free from excess of heat or cold, as adapted in the best possible manner for the enjoyment of a life extended far beyond the
present limits of human existence, and as devoted to peace, to
social and domestic happiness, and to pure and refined enjoyment.

If any new arguments were wanting to prove the insufficiency of human reason in its attempts at remote supposition and
its inability to account minutely for the primeval state of
things, they might be derived from this conjectural history of
the first period of the globe. Whenever Whitehurst is directed
by the arrangement of facts, his ideas are always accurate and
philosophical. But when he enters the field of wild speculation,
his opinions are almost as visionary as those of the elder theorists. His account of the strata of Derbyshire offers an admirable specimen of acute observation and plausible induction. It
is one of the earliest and one of the most perfect of true geological sketches.

On the contrary, his theories of primitive islands and of
the perpendicularity of the axis and orbit of the earth are mere
ingenious dreams. He endeavours indeed to show their conformity
to reason and to sacred history, but without success, for the
oblique rays of the sun in the period of twelve hours could
scarcely possess any effect upon the poles. Ice in consequence
would probably be continually accumulating; there would be no
extensive winds which are produced only by the ascent of columns
of air from heated tracts of land of considerable extent and
which are necessary for blending the parts of the atmosphere together and for equalizing their temperature. The lower strata
of the air would in consequence soon become impure and unfit
for the purposes of life. And on these ideas, allowing the
hypothesis, the antedeluvians, instead of being gently warmed

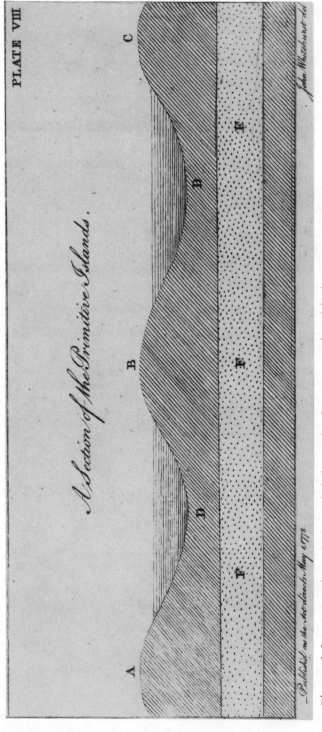

Figure 4.2. A Section of the Primitive Islands. From John Whitehurst, *Inquiry into the Original State and Formation of the Earth*, pl. VIII.

by a perpetual summer, would have been chilled by the frosts of
an eternal winter. Instead of breathing a purer and more ether-
ial fluid, they would have been exposed to an atmosphere filled
with noxious and suffocating vapours. Animated nature would
have fallen rapidly into decay, and there would have been no established principle of reproduction.

The system of Mr. Deluc [1727-1817; DSB 4:27],[6] as it first
made its appearance in 1779, embraces a smaller number of objects
than that of Whitehurst, but his discussions are more minute and
his observations more general and extensive. Whitehurst examined
with attention the facts that were presented to him, but his
sphere of observations was unfortunately limited to a particular
district, and he was unacquainted with the greatest and most important of the mountain chains belonging to the globe. Mr. Deluc,
a native of Switzerland, from his infancy had been accustomed to
the contemplation of the grandest and most elevated of the mountain chains of Europe. He studied these great and interesting
masses with attention and was enabled to compare the geological
phenomena presented by them with similar phenomena in Germany,
France, and England. In his *Letters to the Queen,* facts are recorded with a truly philosophical spirit, and his hypotheses
are generally advanced with diffidence and defended with mod-
eration and candour.

Mr. Deluc commences his enquiries concerning theory at
the Mosaic account of the creation. He assumes a certain
arrangement of things which he calls primordial and which he
considers as resulting from the immediate will of the Almighty.
He supposes that the surface of the globe at its first formation contained a number of irregular and extensive caverns
which were equally extended beneath the bed of the ocean and
below the islands and continents.

He refers the Deluge to the destruction of these caverns
by subterraneous fires, and he supposes that at the time of
this great convulsion of nature the primeval lands were broken
down and covered by the ocean, which rushed into the abyss
that they had overspread.

The present islands and continents which before existed
in the water were in consequence uncovered so that the places
of the land and sea were altogether changed.

Instance

Mr. Deluc conceives that the rocks which contain no organic remains and which form the bases and the summits of our
highest mountains were part of the primordial matter of the
globe. The stratified rocks containing shells, he regards
as formed at the bottom of the antedeluvian ocean; and the
basaltic rocks which so often occur contiguous to the common
stratified rocks, he considers as produced by marine volcanoes.

In his first speculations, the philosopher of Geneva made
no attempts to explain the minute parts of cosmogony, and he
entered upon no discussions concerning the composition of the
chaos or the laws of its arrangement.

In his later writings, he appears, however, to have been reduced by too ardent a desire of generalization. When he considers a confused fluid in which all things were dissolved as the first form of our globe, and light as the principle which separated its parts and gave to them regularity and order, and when he regards this light as different from the solar light, it is impossible to follow with pleasure the wanderings of his imagination; it is impossible not to express a wish that he had rather confined his active and powerful genius to observations and to reasonings upon probabilities than have suffered it to waste its strength in vain attempts to penetrate into mysteries which have been wisely concealed from us, and the knowledge of which, even if it could be obtained, would be comparatively useless.

The system of the earth of Wallerius [1709-1785; DSB 14: 144] agrees in many of its points with that of Mr. Deluc, but it is less valuable as containing more hypotheses and less of observation.[7]

The learned Swedish professor, who followed the steps of the celebrated Linnaeus in attempting to produce an accurate classification of minerals, has not been very successful in his efforts to apply a sound understanding, fitted for the arrangement of facts, to conjecture and analogical conclusions. He founds an important part of this theory upon the same basis as Whitehurst, that of the perpendicularity of the axis of the earth to the plane of its orbit before the Deluge, and he conceives that the inclination of the axis was the first cause of the inundation of the water which broke down and destroyed the land. He reasons much more upon the primordial state of things than upon their present arrangement, and his ideas concerning the first forms of matter are very little connected with established phenomena.

A very interesting account of this theory may be found in Mr. Howard's *Letters on the Scriptural History of the Globe*,[8] a work which, excepting the hypothetical parts, is filled with much useful and curious information and which contains some beautiful views of the alterations of the globe at present taking place.

A short time after the system of Wallerius had been brought forward, *The Theory of the Earth* of Dr. Hutton [1726-1797; DSB 6:577] made its first appearance, and this period may be almost considered as forming an era in the history of geology.[9] Few works have ever excited more attention, few works have been more warmly admired or more vehemently attacked, and it has at least answered the important end of keeping the public interest alive to the facts and discoveries of the science.

Many of the elements of it may be found in the writings of preceding authors, in those of Hooke, of Buffon, and particularly of Lazzaro Moro [1687-1764; DSB 9:531], who supposed all mountains to have been formed by volcanic eruptions.[10]

Dr. Hutton, however, in cases when he has borrowed materials, has given them in such a manner, with the result of his

own observations and hiw own reflections, as to form a whole in
which all the parts are well adapted to each other. Though it
is easy to doubt of the truth of the system it is scarcely possible not to admire the order and ingenuity with which it is
contrived.

Dr. Hutton, reasoning like Buffon, conceived that the agency
of the atmosphere and of water must occasion the mountains of the
earth constantly to diminish, and he supposes that a time must
arrive in the lapse of ages when all the solid land which constitutes our present continents and islands must be wholly destroyed if there did not exist another cause capable of renovating
the land and of opposing the agency of water.

This cause he considers as fire acting under the pressure
of the sea, and he attempts to prove that, in the cycle of events,
the solid earth that we at present inhabit had been formed by
fusion at the bottom of a former ocean and raised from it by
igneous explosion.

Instance

Dr. Hutton attempts to support his system by numerous
appeals to observation and to experience. He quotes a variety
of instances in which rocks have been observed to diminish, in
which the beds of rivers have become deeper, and in which the
ocean has gained upon the land. He details the history of the
elevations of various islands in the midst of the ocean by submarine volcanoes. And assuming the production of basalt by
fire, he attempts to show that there exists a strong analogy
between this substance and granite, and that the mixture of
basalt with the rocks containing organic remains bears a strong
analogy to the intersection of the primordial or primitive rocks
by veins of granite.

The two theories that have been most opposed to that of Dr.
Hutton have been the invention of Professor Werner [1749–1817;
DSB 14:256] and Mr. Kirwan [1733–1812; DSB 7:387].[11] In the
geognosy of Werner, water is made the great agent both of
decomposition and of formation. Werner conceives that in the
beginning of things all the existing substances composing the
globe were in a state of solution by the aqueous fluid. And
he assumes that in consequence of the laws of chemical attraction and in consequence of the polarity of the particles, they
assumed the forms that at present belong to them.

The rocks that are called primordial or primitive and that
contain no organic remains, he supposes the first to have been
deposited from the original aqueous solution, and he considers
them as the result of crystallization and of an arrangement of
matter in which all attractive powers were duly opposed to each
other.

After the first production of rocks from the dissolving
fluid, he supposes counterbalancing affinities in consequence of
which certain substances before dissolved were thrown down not
in the crystallized form but rather in a state of mechanical division, and from their mixture with the substances of primitive
formation he supposes a new class of rocks, which he calls the

transitional rocks, to have been formed. At the period of the deposition of the transitional rocks, Werner supposes the ocean to have been peopled with living beings, and hence he accounts for the shells and remains of marine animals existing in them.

The last epoch in his geognosy is the production of the secondary or highly stratified masses containing in almost all their parts organic remains. These he conjectures to have been produced wholly by the mechanical precipitation of the lowest solid materials which had been suspended in the ocean, and which the former depositions had rendered free so as to enable them to be thrown down in regular strata with the decomposing materials of the animated beings belonging to the sea.

Werner perhaps more than any other man has had opportunities of observing the facts of geology belonging to a particular district in their most minute and abstruse relations, from his situation as Professor of Mineralogy in the School of Mines at Freiberg, the greatest perhaps ever established. He has been enabled to inspect the most important of the forms in which the varieties of subterraneous bodies are presented to us.

And yet in theory he has contented himself with stating what he even considers as a mere history of guesses. He has allowed insurmountable difficulties in the way of a perfect arrangement, and his cosmogony seems designed to satisfy the ardent imagination of students in geology and to awaken the attention of scholars rather than to communicate any accurate views of nature or to satisfy the wishes or desires of the true philosopher.

Mr. Kirwan, in his geological essays, has in many cases followed principles of reasoning similar to those adopted by Werner. Like the philosopher of Freiberg he has supposed a universal aqueous solution of the solid materials of the globe, and he has accounted for their deposition in a similar manner.

But he has in all cases endeavoured to adapt his own particular theories to the sacred history of the creation and in consequence has often exposed himself to the imputation of erroneous interpretation. Thus when he considers the sublime passage, "God said let there be light and there was light," as merely expressing the bursting forth of volcanic fires in the arranging, chaotic mass, it is impossible not to suppose him mistaken. And when he conjectures the sentence which states that the "spirit of God moved upon the face of the waters" merely to imply that a great evaporation took place from the sea, there is no sound reasoner who will not believe the meaning to be forced and unsatisfactory.

Mr. Kirwan has done much for science. As a chemist his labours are always valuable, as a recorder of facts his statements are estimable. In censuring his geological opinions, I censure only what may be considered as the lighter ebullitions of his fancy. His productions on general philosophy are often deep and refined, and always interesting and instructive. His reputation has been well and fully merited. And that he may

long continue to enjoy that reputation must be the wish of
every disinterested friend to the cause of true philosophy.

Controversies between the theorists who consider the present state of the globe as resulting from the agency of fire,
and those who regard it as the effect of the power of water have
been carried on with great ardour. And the two systems have
been named in opposition to each other, the Plutonic and the
Neptunian systems from Pluto and Neptune, the one supposed by
the Greeks to be the tutulary god of the subterranean fire and
the other, the deity of the ocean and of water. The theory of
Dr. Hutton, like that of Werner and of Kirwan, has been principally founded upon conjectures, for as yet there have been no
positive proofs that the principal materials of our globe are
either soluble in water or capable of being fused by fire and
deposited in regular forms in consequence of cooling.

The masterly experiments of Sir James Hall [1761-1832; DSB
6:53] have indeed shown that basalt rendered fluid in our furnaces may be crystallized by a slow diminution of temperature
and that limestone may be fused under great pressure without los-
ing any of the elastic fluid with which it is combined.[12] But
no strong analogies can be applied, from these most interesting
facts, to granite, to the regularly formed gems, or to any of
the perfectly aggregated rocks containing no organic remains.

The general theory is almost as obscure as before, the
early operations of nature are unknown, and the laws of the
formation of the habitable globe are still almost as little
understood as those of the production of the general planetary
system of the sun and of the fixed stars. We are informed by
Clemens Alexandrinus that the Egyptian priests made a statue
of the god Serapis in which they blended together all the known
metals, all the stones, and all the soils commonly found in
Egypt, and this statue has been supposed to represent the primordial state of things. A useful lesson at least may be derived from this narrative. No person uninstructed in the history, supposing the image now existing, would be able to reason
from the form concerning the materials, and no person from view-
ing the materials would be able to develop the means by which
they were mingled or the end for which they were designed.

The powers of man are scanty and definite; even the origin
of certain well-known human productions are far beyond the power
of his investigation. He is unable satisfactorily to explain
the creation of the pyramids of Egypt or of the druidical pillars of Stonehenge. And yet he is sufficiently daring to endeavour to apply his mind to the most recondite operations
which have been produced by divine intelligence, to attempt to
explain the changes that have taken place beneath the foundations of the earth, to examine the great laws of the creation
of the systems of the universe, and to reason concerning their
origin in that infinite space which he is wholly unable to penetrate or even accurately to imagine.

As I have spoken of the theories of Dr. Hutton and Mr.
Werner, it would be unjust if I passed over, without noticing

the latest geological publication, the elucidations of the
Huttonian theory by Professor Playfair [1748-1819; DSB 11:34].[13]
In this excellent work, the ideas of Dr. Hutton are developed
with precision and beauty, and an arrangement is given to them
infinitely superior to that exhibited by the original author.
Dr. Hutton is obscure and perplexed from the multitude of facts
which crowded on his mind. Mr. Playfair, gifted with the faculty
of selection, has discriminated such phenomena only as are calculated to elucidate his opinions. He has given to the Plutonian
theory a new, a more philosophical, and a more fascinating form.
And whatever may be the fate of the opinions of Dr. Hutton, the
work of the distinguished professor of natural philosophy of
Edinburgh considered as a literary production will stand the or-
deal of time and of the most rigid criticism. The elegance and
accuracy of his delineations must be always applauded, and the
precision and clearness of his method and of his style will constantly be subjects of admiration.

In the investigation of geology, as in almost all other
enquiries to which the human mind has been ardently devoted,
very few speculations have indeed been formed not possessed of
some immediate or remote applications to the real progress of
science. The understanding is permanently guided by experience,
and brilliant delusions, even though consecrated by the efforts
of genius, cannot very long continue to deceive the public.

Useful truths are often ascertained in the attempts made
to detect imposing errors, and the appeal to experiment, which
is the last and the only certain test of the merits of opinion,
can hardly fail to lead to discovery. Hypothesis uniformly produces discussion, and the more ingenious and the more active the
talents by which it was formed, the greater is the probability of
a minute and serious examination of facts.

To explain nature and the laws instituted by the Author
of nature and to apply the phenomena presented in the external
world to useful purposes are the great ends of physical investigation, and these ends can only be obtained by the exertion of
all the faculties of the mind. And the imagination, the memory,
and the reason are perhaps equally essential to the development
of great and important truths.

Lecture Five

Almost all the later hypotheses in cosmogony noticed in the former lectures have in some manner led to the acquisition of facts and to the observation of nature. The defenders of them have sought for phenomena subservient to their own particular views; the opposers of them have endeavoured to discover appearances incapable of being reconciled to the theory. The real knowledge of the earth has been extended by the efforts of both, and practical information has thus arisen from speculative research.

The objects connected with the development of a perfect chemical theory of the globe are almost innumerable. Time and varied labour and the application of great ingenuity will be necessary for a general elucidation of causes; the foundations of the science however are laid, and the first principles of geology are sufficiently distinct and intelligible, those principles which relate to the properties and nature of the great masses which compose the known part of the surface, the powers by which they are modified, and the order in which they are maintained.

When the general exterior of the earth is considered, the appearances are numerous and diversified. Hills and vales are seen green with vegetation. Mountains hide their summits in the clouds, and the light blue expanse of water contrasts with the dark tints of land. The view is warm with beauty and animation.

On penetrating beneath the surface, however, a new and altogether different order of things is presented to our attention, all is dark and silent. Soils, earthly strata, and rocks are blended together without regularity of form. Veins of

metallic ores and of crystallized stones intersect them in different directions, and by superficial examination deformity and
disorder only are perceptible in the subterraneous mineral kingdom.

This effect of confusion, however, is but transient upon
the investigating mind, for by scientific methods of comparison,
a distinct arrangement is perceived even in those rudest of the
forms of matter. An analogy between their parts is ascertained
and their utility and importance in the existing series of natural events appear great and obvious.

I mentioned in a former lecture that there is a great resemblance in the mineral production of extensive districts situated at distances from each other. Similar soils and earths
are usually discovered incumbent upon similar stony strata,
and the various rocks that occur together in one position are
likewise generally found together in other positions. Thus
the higher mountains of France and of Spain are similar in
their nature to those of Britain. Most of the specimens brought
from America presented appearances of the same kind as those
found in Europe. And all the rocks and stones that have been
as yet procured from New Holland,[1] a country containing so many
novelties in botany and zoology, have not any of the characters of new mineralogical species and precisely resemble those
of the old continent.

It is on this circumstance principally that our method
of classing substances according to their geological relations
is founded. It is this circumstance which gives a simplicity
and a unity to the arrangement of the science and which enables
us to judge of the position and nature of great masses from the
examination of a few of their minuter parts.

The distinctions between the primitive and the secondary
rocks have been already often noticed, and an attention to the
difference of their natures is of the highest importance in
every part of the study of geology. All substances that contain organic remains that bear evidence of having been produced
since the existence of living beings on the globe are excluded
from the series of primitive masses, as likewise all strata
fused by volcanoes or deposited from alluvion by the late agency
of water.

The primitive rocks are the substances which occupy the
lowest part of the surface of the earth penetrated by human
labour. And they form the summits of most of the highest eminences of mountainous countries.

The great masses of primitive rocks are of several species
which have characters peculiar to themselves and wholly different from those of all other substances. These characters it will
be necessary for us particularly to attend to.

They may be considered geologically as of two kinds. The
first are the external characters of particular specimens considered as to appearance, weight, and hardness. The second are
the characters of the rocks considered as masses in the natural
connection, consisting of general form, of fracture, and of as-

pect. The characters gained from minute inspection of single
specimens are the most accurate, but the general outline and
shades of rocks, though not uniformly to be trusted, do offer
very useful indications.

I shall endeavour to illustrate the first class of characters by the exhibition of pieces of rocks, and the second by
means of coloured sketches. These illustrations will I fear
be the most tedious of the course, for accurate knowledge can
only be gained by minute inspection, and in a room so large as
this, it is impossible that all the appearances can be distinctly
observed. I shall however use every effort to describe the properties in such a manner that they may be easily remembered, and
by those who are desirous of acquiring practical information, the
specimens may be afterwards examined accurately and in detail.

One of the most common species of the primitive rocks is
granite. The word signifies granulated, and when superficially
examined the stone appears as a number of small fragments cemented together. It is, however, in fact composed of three
bodies confusedly crystallized together. They are named quartz,
feldspar, and mica.

Hardness-Crystallization-Lustre

The differences of the colours of granite are principally owing
to the feldspar and mica. The feldspar is found in shades of
yellow and brown. It is sometimes green and often red. The
granite of Pompey's Pillar is red. Granite sometimes contains
another constituent part, which is schorl.[2]

Instance

And the mica is often so small in quantity as to be scarcely
perceptible, and in this case the rock appears as a mixture
of crystals of feldspar and quartz.

Instance

Graphic granite-Pliny pyrope-[Illegible]

Granite, as to its texture, is one of the firmest of stones
and is in general the least liable to decomposition and decay.
Most of the great edifices of antiquity that have been least
altered in their forms have been built of this substance, and
its durability is no less equally attested to by the great
masses of the pyramids of Egypt than by some of the minute works
of elder sculpture of that celebrated country lately brought into
England, and which are wholly unimpaired by time.

Granitic rocks are usually found in large blocks accumulated together without regularity of form and possessing un-
equal edges. They occur, however, sometimes in layers, when
they bear the name of gneiss, and in this case the plates of
mica appear interwoven upon extensive surfaces which are arranged
at intervals between the blended feldspar and the quartz.

In this picture, the common aspect of granite is exhibited
in the representations of the pile of rocks marked No. 1 [see
endpapers].

The sketches imitate the position of granite rocks in
the island of Mull[3] where the rocks are comparatively
small.

In this picture, which has been before exhibited, the lower rocks, very much inclined to the horizon, may be conceived to represent the granite in layers, or gneiss. All the other rocks represent the substance in blocks which are distinct and placed on each other. And their distinctness is exceedingly evident in the rock represented by this mass. It is called the Logan Rock.[4] It is an isolated body of nearly 40 feet in circumference and so exactly formed on the point of another rock that it can be easily moved to and fro by the hand, though the united force of 100 men would be wholly insufficient to move it.

Figure 5.1. Logan Rock. Although the setting in this drawing is not accurate, the rectangular blocks of weathered granite are well illustrated, and Logan Rock is shown here as Davy described it, balanced on the tip of another stone. At the right, a person leans against the balanced rock. From Richard Polwhele, *The History of Cornwall,* new ed., 7 vols. in two (London, 1803-10), 4:131.

The aspect of granite is well exposed in every part of this view. No rock is grander in form nor more sublime in structure. Placed at the last extremity of our land, it seems well adapted from its magnitude, solidity, and strength to resist the force of the waves of the Atlantic and to prevent the encroachment of the ocean upon the land.

Syenite, another of the primitive rocks, derives its name from Syene in Egypt. When examined in small masses it appears very similar to granite. One of its ingredients is the same, the feldspar, but it contains neither quartz nor mica but hornblende.

Instance
Hornblende harder than mica
Dark crystals

The aspect of syenite when presented in great masses in
nature is very peculiar. It neither occurs stratified nor separated in blocks, but it presents a number of irregular fractures
and pointed crags and its colour is generally grey or yellow.
The figure marked "No. 2" in this picture represents syenite, and
several of the monuments in the British Museum are of syenite.
Both granite and syenite are composed of distinct crystals, but
porphyry, which is often found near them or incumbent upon them,
is very different in its texture. It consists of crystals of
feldspar imbedded in a mass of hornstone or of jasper.

Porphyry is found in a great variety of colours. Of the
most beautiful kind is the red porphyry, and of this there is a
large pillar in the British Museum. The aspect presented by
porphyry in some measure resembles that of granite, but the
blocks are smaller and the surface is almost always smooth.

Instance

The masses of porphyry are rarely large and hence insulated rocks or cliffs of it seldom present much majesty of form
or variety of appearances. But when blended with wood and river
scenery, as it is in several valleys near Ben Nevis in Scotland,
the appearance is exceedingly beautiful and its high tints increase the effects of the green of vegetation and the full light
off the water.

Another primitive rock which occurs almost as frequently
as granite and much more frequently than either porphyry or
syenite is micaceous schist. It consists principally of mica,
whence its name, but it likewise contains quartz. In its appearance it is one of the most splendid of rocks. It occurs generally
in irregular layers which have the appearance of leaves.

Instance

There are two other species of primitive schist, for the
name has been applied to several stones arranged in layers.
These are rocks of a simple constitution. They are both opaque
and similar as to aspect, but the one is so hard as not to be
cut with a knife and the other is exceedingly soft.

Instance

The hard is named primitive siliceous schist and the other
argillaceous schist. All the primitive schists in the exterior of their form present considerable irregularities and
their layers break in points, in consequence of which when
masses are large their outline is often very bold and impressive.

Serpentine is one of the primitive rocks which occurs
least frequently in nature. It is simple in its constitution
and is one of the most singular of mineral productions. It
is smooth to the touch, is scarcely scratched with the knife,
and has a peculiar lustre. Its colours are generally mixed and
principally red and green which are diffused in streaks similar
to those on the skin of the serpent, whence its name. By the
ancients it was called "ophite" and was much-used in ornamental

works. Pliny, who has given a detailed account of this stone, attributes to it many wonderful virtues, such as that of healing the bites of serpents, and he recommended a piece of it worn suspended from the neck as an effectual guard against every species of enchantment.

Serpentine at a considerable distance is easily distinguished by its aspect from all other rocks. It occurs in masses which generally approach to the square figure and it often presents a number of small and irregular chasms.

Instance

An attempt has been made to delineate the aspect of serpentine in this group of rocks, No. 5.

Serpentine is one of the primitive rocks which contains the greatest number of perforations and of orifices, and when it occurs upon a great scale its appearances are in the highest degree picturesque.

In the large masses of serpentine as they exist in nature, nothing can exceed the variety of the colours and the smoothness and polish of the surface. Red, dark green, brown, and yellow all appear, sometimes distinct, sometimes softening into each other. The white foam of the wave becomes a cause of contrast, and if granite is the most sublime of the primitive rocks in its forms, serpentine may with propriety be said to possess the highest character of beauty.

Quartz rock[5] forming immense masses occurs in some primitive districts. It is granular in its fracture and sometimes mixed with small quantities of mica. In its aspect it appears in layers or in imperfect blocks having four-sided forms.

Instance

This quartz mountain is of a pure white and when it is first considered at a distance, it appears amongst the other mountains as if clothed in perpetual snows. The curved lines of the strata that compose it are very beautiful, and when illuminated by the sun, the white light from it produces the highest effect of brilliancy and splendour.

It was once doubted whether any limestone existed free from shells, but the most accurate researches have lately demonstrated that there are undoubtedly primitive rocks of this substance found contiguous to the other primitive rocks.

Primitive marble differs from common marble not only in being free from organic remains but likewise in its fracture, which is soft and grained or like that of loaf sugar, and it is composed of minute crystals of calcareous spar. Its colours are various. It is so soft as to be scratched with a knife and it effervesces copiously when thrown into vinegar.

Instance

The statuary marble of the ancients is primitive limestone.

The aspect of white primitive limestone is somewhat similar to that of quartz rock but its lustre is less considerable.

Instance

It is almost always arranged in strata; it seldom presents a rough outline. Its forms are curved and its declivities gentle.[6]

These are the principal rocks that form the solid foundations of the surface of the earth. There are many varieties of
them, and there are some other stony substances that occur mixed
with them. But I have thought it necessary in this lecture to
attend only to the nature of the great principal masses and to
endeavour to fix your attention upon their decided characters.
On a future occasion I shall fully examine the other primitive
substances, the stones that exist blended with the primitive
rocks either in beds or veins, the metallic ores that they contain, and their indications.

The primitive rocks in their position with regard to each
other seldom occur precisely in the same arrangement. Granite
is the rock usually found at the greatest depth and it is generally connected with micaceous schist, syenite, or with quartz.
Serpentine, schist, and granular limestone often accompany each
other, but serpentine and granular limestone seldom occur either
incumbent upon, or alternating with, granite.

Thus in Cornwall, which is most distinctly a primitive
country, the granite and micaceous and siliceous schist meet
and form the principal rocky basis of the land. The limestone
occurs only in alternation with or upon the schist, and the serpentine district is bounded on one side by micaceous schist and
on the other side by syenite.

Instance

In Scotland, in the great mountain chain of the western
highlands, of which a part is represented in the background of
this picture, the highest mountain in the island, Ben Nevis--
which rises 4500 feet--is micaceous schist. The next marked is
quartz rock, and granite and porphyry form the other mountains.
The serpentine and marble in Scotland are usually found low and
at the bottom of hills of micaceous schist or syenite.

There is a similar arrangement in other great mountain
chains. Thus in the highest eminence of the Alps, which is
represented in this sketch, the principal rocks are syenite,
granite, and micaceous schist. Granite forms the great basis
of Mont Blanc.

A number of geological descriptions which demonstrate
the same facts may be found in the writings of Mr. Deluc and
of de Saussure [1740-1799; DSB 12:119], but it will be unnecessary for me to multiply the instances as they are in most cases
analogous.[7] All the details of M. de Saussure upon the Alps are,
however, in the highest degree worthy of being pursued on account
of their relations to the structure of the globe. This excellent
man constantly studied nature with the views of a true philosopher. He seldom indulged in speculation. His theories are
few and founded wholly upon his own observations, and his nar-
ratives are composed with a simple and unaffected eloquence
which proves that he taught rather to communicate knowledge
than to excite admiration.

The different primitive rocks have very different modes of
junction and of transition into each other. In some cases the
parts of their union are in right lines, a circumstance which

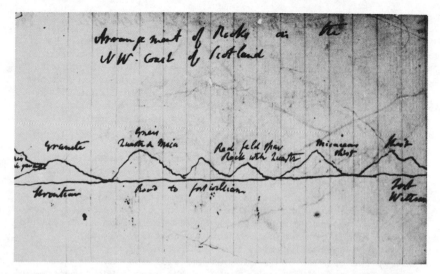

Figure 5.2.   Arrangement of Rocks on the Northwest Coast of
Scotland.   Davy's drawing represents about thirty kilometers of
the coast, from Strontian at the southwest to Fort William
toward the northeast.   Davy MSS (15e, 138) 1804.

Figure 5.3.   Ben Nevis.   Davy MSS (15e, 137) 1804.

often occurs in the meeting of siliceous schist, and granite, and porphyry.

Instance

Serpentine, syenite, and micaceous schist sometimes pass into each other by very gentle gradations.

Instance--Lizard

And where granular limestone is placed upon micaceous schist, near the point of junction it uniformly contains larger or smaller proportions of mica.

Instance

Granite and micaceous schist sometimes have their parts mixed at the point of junction and in other cases they inosculate, sending forth distinct ramifications of veins.

Instance

This fact occurs on a considerable scale at St. Michael's Mount in Cornwall.[8] The great body of the mountain is of granite, but at its base a stratum of the schist occurs, and thin layers of granite may be traced to a considerable extent running into the schist, whilst ramifications of the schist in many places penetrate into the granite.

Instance

Those inosculations of granite and schist which occur abundantly in different parts of the world have been subjects of great disputes amongst the various theorists. Werner and most of his disciples, considering them as veins of granite, attempt to show that they are granite of a different kind from that which forms the first of the rocks deposited by crystallization and that they are of a secondary order and formed by a much later infiltration of water.

To the Huttonians they offer a strong argument in favour of the formation of granite by igneous fusion. The rock in which a vein is found, says Dr. Hutton, must have been prior in existence to the vein itself; therefore the granite must have been formed after the schist. And as the chasm in the schist has been completely filled by the granite, this substance must have been fluid, and there exists no agent but fire adequate to its fusion and to its projection into the schist.

Neither the reasonings adopted in the one theory or in the other will, I conceive, carry conviction to the cool and unprejudiced observer. The granite veins (if they be allowed the name) are almost always similar in their appearance to the granite rock with which they are connected, and this cannot even by the Wernerians be called with any plausibility a secondary rock.

With regard to Dr. Hutton's opinion, it is true that where single and extensive insulated veins occur in large rocks the rock must be considered as primary in its formation when compared with the vein. But in a case where two rocks composed of very similar materials merely blend by ramification, it is not necessary to suppose them of different periods of formation, for there are many cases of crystallization in which different salts that solidify at the same time from their solution, inosculate with

each other and produce effects not dissimilar to those observed
in the mixture of the primitive rocks.

On the extensive ground of opinion I confess that the general
structure and appearances of the primitive rocks offer very few
facts favourable to either hypothesis. For as I have already
said, their principal materials are neither fusible by fire nor
soluble in water. As yet no particle of pure quartz has been
rendered fluid by the strongest heat of our furnaces assisted
by the most powerful effects of air. It is therefore scarcely
possible to believe that mountains of this stone more than a
mile in perpendicular height and many miles in circumference
can have been rendered fluid by subterraneous heat excited
without the access of air and supposed to act even immediately
beneath the water of the ocean.

There is indeed another strong argument against the primitive rocks having been formed by fire from the remains of
former continents and islands. Coal or carbonaceous matter is
one of the substances very abundant upon the surface, and in its
pure form it is an element incapable of being produced and is
not, in the slightest degree, volatile by heat. And it ought
therefore (assuming the probability of the theory of Dr. Hutton)
to be found in or amongst the groups of primitive substances,
with granite, with micaceous schist, and with serpentine and
granular limestone, but this in fact is not the case. And in
almost all the mountain chains, scarcely any vestiges of coaly
matter occur and in no part do they exhibit any marks of having
been produced from the ruins of a former system containing
organized beings and fitted for the purposes of life.[9]

If the Huttonian or Plutonic theory of the formation of
primitive rocks be founded upon conjectures, the Neptunian geognosy of Werner and Kirwan is almost wholly built upon error
and chemical impossibilities. Thus the supporters of it suppose
stones and metals soluble in water, and yet precipitate from it
without any change in the fluid, and they make the same agent
at once the dissolving menstruum and the crystallizing power.

Whenever natural causes can be fairly investigated, it
is the business of philosophy to endeavour to trace them. And
no object of research, however hidden, connected with the discovery of the laws of nature ought to be neglected provided our
instruments of investigation can be applied to them, provided
they are capable of being elucidated by analogical inferences
from known facts. But in cases where there are no histories
to guide us, no distinct reasonings to assist us, and no experiments to enlighten us, there the human powers must be applied in
vain, and all the efforts of ingenuity wholly wasted. The limits
between what is capable of being known and what must be forever
concealed from us with regard to the theory of the earth are not
perhaps very distinct. But one of the first opinions and one of
the most simple, with regard to the subject of the present discussion, is perhaps the one that will finally be adopted by
enlightened observers.

Figure 5.4. The Four Strata out of the Order of Specific
Gravity, Teesdale. Davy's note on this drawing refers to the
view held by some Neptunians that rocks precipitated from the
primordial ocean in the order of their specific gravity, a
process that would have positioned the densest rocks on the
bottom and the lightest on top. He provided no further comment
on the adjacent pages of his notebook. Davy MSS (22a, 130) 1804.

When the word "primitive" was first applied, which I believe was by Lehman [1719-67; DSB 8:146],[10] it signified that
matter of our globe as yet unchanged by any known natural operations. In this sense, its meaning was definite and it was sufficient for all the purposes of science. In every system some
primordial state of things must be allowed. And where we can
perceive no certain indication of a prior arrangement, there it
is reasonable for us to rest, there it is reasonable for us to
fix the foundations of our science.

When Newton applied his powerful genius to the development
of the laws of the planetary system, he did not begin his researches by endeavouring to develop how they received their present forms or how they were endowed with powers of motion. With
true sagacity he left those questions untouched, and instead of
indulging in conjectures and exposing himself to the censure of
being a vague theorist, he was contented with the arrangement of
known phenomena and the explanation of nature by analogy compared
with facts. And in consequence he had the glory of having

discovered some of the most extensive and most sublime of physical truths.

Such an example may be successfully imitated in all researches. In the study of the chemical history of the earth, the first object of enquiry ought to be the present state of things. When this is accurately understood just deductions of reason will necessarily arise with regard to the past and the future. If Dr. Hutton and Mr. Werner had contented themselves with explaining those operations by which our globe is constantly preserved as a habitable world in the series of events at present taking place, they would have deserved a much higher praise than that of having been the founders of ingenious hypotheses. But when attempting to explain appearances, they attribute to agents powers which they have never been observed to exert, or refer effects to causes, the operation of which they are ignorant, their suppositions do not merit the name of science. Such speculations however brilliant will pass away and be forgotten, whilst the facts that have been connected with them will remain and become the parts of a more permanent theory and of one more conformable to the true order of nature.

In the present state of our knowledge, we are wholly unable to explain the origin of primitive rocks, but many interesting and attainable objects are presented to us in the study of their composition, of the agencies to which their present state is owing, and of their general uses in the economy of nature.

The appearances of primitive rocks are almost infinitely diversified and yet their constituent parts are comparatively few. By the refined methods of chemistry, even the hardest stones are capable of being brought into solution and their elements separated from each other. This fluid contains the matter of quartz rock dissolved in an alkaline lixivium, and by means of an acid, the earth that it contains may be separated in a finely divided state. By other analogous methods which were exhibited in the autumal course of lectures on chemical analysis, the constituent principles of other rocks may be separated from each other and obtained in a form of purity.

By such experimental processes it is demonstrated that almost all the varieties of the primitive rocks that have been described consist chiefly of different combinations of four elementary earths, called silex, alumina, lime, and magnesia, and a small quantity of iron combined with oxygen, or pure air.

Alumina in its pure and crystallized form exists in the sapphire, siliceous earth in rock crystal. And caustic magnesia, and oxide of iron is its colouring nature. In hornblende and schorl, the quantity of iron is greater. In serpentine, denominations.

In feldspar, the siliceous earth is combined with small quantities of the other primitive earths, and when coloured, it owes its tint to the combined oxide of iron. Mica principally contains siliceous earth and aluminous earth with a little magnesia, and oxide of iron is its colouring nature. In hornblende and schorl, the quantity of iron is greater. In serpentine,

there is a considerable proportion of magnesia united to the
other earths. But white limestone consists of pure lime united
to an elastic fluid called carbonic acid or fixed air, whence
its effervescence when acted on by an acid, and the coloured
limestone owe their tints to oxide of iron.

The elementary earths that form the constituent parts of
rocks not only differ from each other in their general properties, but likewise in the chemical attraction by which they
are united. And hence indications of the composition of stones
may in most cases be gained from their external properties. Thus
the crystallized stones that principally contain alumina and
silex, substances that have a very strong affinity for each
other, are always very hard and difficult to be broken; and the
greater their hardness, the greater the proportion of alumina
that may be expected in them. The stones that contain magnesia with alumina and silex, on the contrary, are usually
soft, as magnesia has no attraction for silex and only a slight
one for alumina. And limestone, which is the softest of all
the primitive rocks, contains lime united by a weak power of
combination to an elastic fluid.

At this time I can only mention these objects; information concerning them may be found in most of the elementary
books of chemistry. They are connected with geology but are
not proper subjects for extensive discussion in a geological
course. The powers of decomposition in the possession of chemistry are very extensive but those of composition are exceedingly limited. As yet none of the arrangements of the primitive materials of the globe have been united by human invention.
We recognize the laws of attraction by which they exist, but
we are wholly incapable of explaining the manner in which
their principles combine. And this part of the enquiry concerning their essence is as yet hidden under the same veil
that conceals the minute facts concerning their origin.[11]

The primitive rocks and mountains, the structure of which
has been described, often occur in chains which extend for
many hundred miles, and their geographical position is of considerable importance in the general order of natural events.

The heat of the air is derived from the agency of the rays
of the sun upon the surface of the earth, and it diminishes
in proportion to height so that the atmosphere at 3 or 4 miles
high is intensely cold. Hence the temperature of mountains is
always much lower than that of plains; and in all climates of
the globe when mountains are above 15,000 feet high, their tops
are covered with perpetual snow. The height of the point of
congelation diminishes in a regular progression from the equator toward the poles so that, in our latitude, it is about 5,000
feet--Ben Nevis being very little within the limits of constant
ice.

It is in consequence of this circumstance that mountain
chains affect in a high degree the temperature of the climate
in which they exist. In the summer, currents of cold air are
continually flowing down their sides, which mitigate the burn-

ing heat of the seasons. In winter they influence the course
of the winds so as, in many cases, to protect the plains from
their effects.

In Britain, for instance, the principal mountains occur
in the north and northeast part of the island. And our hottest
winds in summer and our coldest winds in winter blow from those
quarters so that their situation is well adapted to diminish the
heat in the one case, and to break the force of the cold blast
in the other case. North America, as is well known, is much
colder in winter and hotter in summer than the parts of Europe
under the same parallels of latitude. And this seems in great
measure to be owing to the inferiority of its mountains which
are few in number and none of which can be compared in height
to the Pyrenees or the Norwegian Alps and much less to the Alps
of Switzerland.

In South America, the great chain of the Andes, though
partly under the equator, considerably tends to diminish the
temperature of the climate and to render it habitable. Upon
a smaller scale it may be said that the most fertile of the
tropical islands and those the most luxuriant in vegetation are
such as present the most mountainous aspect.

The effects of mountains upon the temperature of climates
are extensive and they possess other agencies perhaps still more
important. In consequence of their low temperature, they are
constantly precipitating moisture from the air or arresting the
clouds in their progress, or from the thawing of snow upon them,
they are sources of springs and rivers. And the streams pro-
ceeding from them, which at their origin flow only over bare
rocks and water only moss or the lichen, become in their progress the cause of the fertility of the rich valley and produce
the verdure of the cultivated plain.

Even those strata which have no elevation and which are
buried beneath the surface act in preserving the consistency of
the soil, and they prevent water and the matter capable of being
organized from penetrating to too great a depth in the earth
where they would be placed out of the reach of plants and de-
prived of the influence of the atmosphere and of light.

By irregularities in the exterior of the earth, the whole
quantity of surface is considerably increased. And those parts
of it which are not covered with living beings are still in
some measure subservient to their existence and to their enjoyment.

Lecture Six

That a number of rocks and stony substances exist containing forms precisely similar to those of certain parts of organized beings did not escape the attention of some of the earliest philosophers who are upon record as enquirers concerning nature. Xenophanes, who lived nearly 520 years before Christ and who was the founder of the Eleatic sect in Greece, has noticed that shells and bones of fishes are imbedded in some of the marble quarries of Sicily, and he endeavours to show from this fact that the island must have been at some time covered by the sea.

Herodotus likewise notices this appearance and states that shells are contained in considerable quantities in the mountains of Egypt. And from the general saltness of the soil, the brackish taste of the water, and the corrosion of buildings by saline emanation, he conceives that this celebrated country had been produced by the gradual diminution of the ocean.

The opinion of the change of land into sea and of sea into land was indeed current amongst most of the authors of antiquity, both Greek and Roman, who have written upon or alluded to the revolutions that the earth had undergone in past times. And the idea did not even escape the notice of the poets. Lucretius seems to allude to it in his second book when he speaks of the various forms of shells. And Ovid in the fifteenth book of *Metamorphoses* expresses the notion distinctly in some lines of which I shall read a translation.

> Remains that to the waters owe their birth
> Occur in rocks beneath the solid earth;

        Where our green lands their varied face display,
        Once in proud triumph flowed the azure sea.
        And in the change of things and lapse of time
        The conquering waves have gained another clime,
        And where another land its verdure spread
        Is now the moving ocean's troubled bed.

    Strabo, in his *Geography,* has described some of the opinions
of different philosophers with regard to the shells and marine
relics found in the interiors of various continents and islands.
And Pliny has given an account of the properties of many of
these singular substances in his *Natural History,* but as usual
he has blended his details with a considerable portion of the
marvellous. He has described fossil oyster shells which he tells
us are certain cures for lameness and for blindness. He mentions
the shell called the "Cornu Ammonis" and the "glossopetra," but
he considers them rather as productions of the heavens than of
the earth and states that the last substance falls from the
clouds at the time that the moon is in its wane.
    The opinions of Pliny are sufficiently ridiculous with
regard to the organic remains found in stones, but those of
some of the writers in the Middle Ages in different parts of
Europe are, if it be possible, still more absurd. Albertus of
Cologne, for instance, in a work published in the thirteenth
century upon minerals mentions several petrifactions of natural
bodies and particularly one of the branch of a tree. He supposes the effect to have had no connection with the prior
existence of a vegetable, but attributes it to a certain forming power in nature which operates beneath the surface so as to
produce imperfect resemblances of living beings, which like
their archetypes are liable to increase and to decay.
    The opinions that stones grew like plants and animals
and assumed sometimes similar forms was prevalent even in the
four following centuries. Thus Dr. Plot [1640-1696; DSB 11:40]
in his *History of Oxfordshire* attributes all the apparent animal and vegetable remains found in that county to a plastic
energy constantly acting in the bosom of the earth and evolving
new forms.[1] And Sir Thomas Browne [1605-1682; DSB 2:522] supposed that all stones arose from seeds in the same manner as
plants.[2] He even attributed to them a kind of imperfect vegetable life. He ridiculed the idea of their being in many
cases primary or permanent substances, and endeavoured to de-
cry the notion as a popular prejudice. The statement is the
more singular as it occurs in a work written professedly
against vulgar errors and which in other parts often displays
much strength of thought and acuteness of opinion.
    Lhwyd [1660-1709; DSB 8:307] combated the idea of the plas-
tic power but he advanced a notion not much more correct, namely
that organic fossils were formed either from the spawn of
fishes or from organized seeds, raised with vapour from the sea,
and conveyed by rain through crevices into the depths of the
earth.[3]

Hooke, whose name I have before mentioned in these lectures and whose labours can scarcely be too much admired, was one of the first persons who fully demonstrated that the various forms of shells of fishes and of plants were truly organic remains and the relics of living beings that at a former period peopled the globe. In his work upon earthquakes[4] his opinions are fully developed and they were adopted by Ray [1627-1705; DSB 11:313], by Woodward, and the other celebrated naturalists of his age.[5] No sound or forcible objections have been made to them. They have been established by time and are equally comparable to philosophical research and to common observations. Hooke has given delineations and descriptions of various remains of marine animals, but in his age, chemistry and mineralogy had made very little progress, and he was unacquainted with the precise nature and relations of the rocks in which those remains occurred.

This knowledge indeed has been gained very lately and then only by the united efforts of geologists and of analytical experimenters. It is within the last thirty years that accurate methods of classing different secondary strata of the surface have been developed by Wallerius, Deluc, Dolomieu [1750-1801; DSB 4:149],[6] and other celebrated men, and that the intimate natures of the various organic and stony substances they contain have been ascertained by the labours and ingenuity of Pott [1692-1777; DSB 11:109], Bergman, Scheele [1742-86; DSB 11:143], Klaproth [1743-1817; DSB 7:394],[7] and other excellent chemists in different parts of Europe, and that by processes which have been constantly approaching nearly to perfection.

The secondary rocks and strata differ from the primitive in a number of respects. They are usually arranged in beds which, though often composed in their present forms, generally exhibit evidences of having been originally deposited in horizontal layers. Their elevation is in general much less considerable than that of the primitive rocks as they seldom rise more than 3,000 feet above the level of the sea and, I believe, never exceed 8,000 feet. The highest secondary mountain in Britain is Ingleborough in Yorkshire which is more than 3,500 feet above the waters in the Irish Channel. And the greatest elevation of rocks containing organic remains in the Alps is about 7,300 feet, and Mr. Deluc found a petrified *cornu ammonis* at this height in Mount Grenier.

The hills containing secondary strata are usually less abrupt in their forms and more rounded and regular in their outlines than those composed of primitive rocks. A very good idea of the difference of their aspects may be formed from the examination of a view of the country round the Lake of Geneva. This painting represents the view, and it is designed from an engraving attached to *Voyage dans les Alps*.[8]

Instance

The contrast is striking. The regularity and beauty of the stratified rocks covered with vegetation in the real scene must considerably enhance the majestic effect of the great chain of the Alps and must increase the sublimity of the appearance of

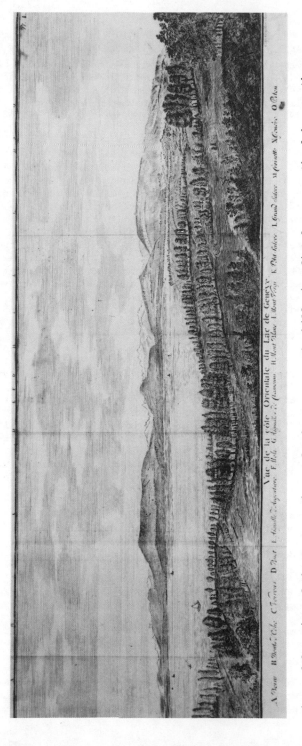

Vue de la côte Orientale du Lac de Genève.

A Nant B Monthey Che C Torrens D Dent E Aiguille F Aiguilles G Aiguilles H Mont Blanc I Mont Tout K P'tit Salève L Grand Salève M Perwette N Genève O Bleu

Figure 6.1. View of the Eastern Shore of Lake Geneva. The hills immediately across the lake are those composed of the Secondary rocks, while the sharp peaks in the background are the Primitive mountains, including Mt. Blanc under the letter "H" which appears at the top edge of the drawing. From H. B. de Saussure, *Voyages dans les Alpes*, 4 vols. (Neuchatel, 1779-96), 1:3. See the endpapers for Davy's later representation of the distinction between the appearances of Primitive and Secondary mountains.

the bare and confused crags and snowy summits of the primitive
mountains.

All the secondary mountains, the interiors of which have
been fully investigated, are found to rest upon the primitive
rocks, and the mode of their union appears very different in
different cases. In some instances when secondary limestone
occurs incumbent upon siliceous schist, the two rocks are merely
closely cemented together without any blending of their parts.
And this happens in many points of the union between the primitive rocks of the north of Devonshire and the strata containing
organic remains that form the whole of Somersetshire and a part
of South Devon. One of the most interesting of the geological
facts of this kind occurs in a cliff a few miles to the west of
Minehead.

Instance

When soft primitive rock as micaceous schist, however, is
contiguous to secondary rocks, parts of it are sometimes found
embedded in them. Thus mica and quartz often appear mixed with
shell limestone near the point of junction. And when compact
marble containing organic remains is found upon granular marble,
there is often an intimate blending of the two rocks about the
point of union. This last circumstance is said to occur sometimes in the Saxon mines.

Upon these general facts of the intermixture of the primitive and secondary substances, Werner has endeavoured to establish a distinct class of strata under the name of "transition
rocks." These rocks, however, occur so seldom and, in general,
in masses of so small a magnitude as to render the distinction
almost unnecessary. And it is more than probable that the acute
mineralogist of Freiberg would not have adopted it himself had
it not been connected with his peculiar theory of the aqueous
origin of all mountains, in which as the primitive rocks were
supposed the results of crystallizations, and the secondary
rocks of mechanical deposition, it was necessary to suppose an
intermediate series produced from their joint effects.

The secondary strata laid in their various positions upon
the primitive rocks occupy by far the largest portion of the
solid surface on which our soils are immediately incumbent.
Their properties, their nature, and their arrangement are objects in the highest degree worthy of a minute investigation.
In many of their parts they abound with metallic veins. They
are the beds in which mineral coal occurs, and they contain a
number of substances subservient to the uses and comforts of man.

To the enquirer concerning natural causes they are perhaps
still more interesting than the masses of primitive origin. They
are monuments of the great changes that the globe has undergone.
They exhibit indubitable evidences of a former order of things
and of a great destruction and renovation of living beings. The
facts that they present are obscure but they bear many analogies
to existing phenomena. The connection between their causes and
effects is mysterious but apparently within the reach of our
faculties, and it is displayed in characters which can be de-

ciphered only with difficulty, but which express sublime truths.
I shall devote the remainder of the time alotted to this sit-
ting to the description of some of the most important secondary
strata and to the discussion of their general arrangement.

I had an intention of entering upon the enquiry concerning
their origin and the great revolution in nature by which they
were formed. But upon further consideration I am inclined to
believe that this subject will be best examined after the general facts have been stated in detail. It is very comprehensive
and important. To be well understood, it ought to be fully developed, and it will most properly form the object of the next
lecture.

The secondary strata differ in their arrangement and position in different countries, but there is an analogy in their
composition and in their properties which demonstrates that they
are in all cases congenerous, of one family, and produced by
similar operations.

Compact limestone or shell marble is one of the most common of the secondary rocks. Even when it is examined in the
smallest masses, its character appears perfectly distinct from
those of primitive marble, and though its composition is nearly
the same, its texture is wholly different. It presents either a
smooth surface when broken or a fracture exhibiting a number of
small plates and has no appearance of being composed of distinct
crystals.

Instance

Secondary limestone is of many kinds, and though its principal constituent parts are uniformly lime and carbonic acid, yet
it often contains other earthy matter, particularly magnesia.
The purest limestones are those which most rapidly dissolve in
acids and which leave the smallest quantity of insoluble matter.
The limestone that contains magnesia is very difficultly soluble.

Instance

The pure limestone is that best fitted for the purpose of
manure and for improving lands, but the magnesia limestone in its
calcined state is pernicious to vegetation, though it makes an
excellent cement which becomes solid and hard under water.

The brown and impure limestones that contain much iron and
other foreign matters are likewise generally much more proper for
calcination and application to the purposes of building than the
white and semitransparent marbles.

Chalk is a secondary substance of the same composition as
limestone, and it differs from it principally in the form of its
aggregation. It consists of a number of particles loosely adhering to each other and not united by chemical attraction.

Instance

The different limestone and calcareous rocks often occur
in different strata in the same district and they are no less
distinguished by external characters than by their containing different organic remains. Thus the shells found in the
light grey and yellow strata of pure limestone are very different from those in the magnesian limestone, and the magnesian

limestone contains organic substances very different from those
found in chalk.

Instance

Most of these shells and remains are in some measure similar
to those of species at present existing, but in few cases, I believe, [are they] perfectly identical, which proves that there
has been a great change in the generations of the beings that
produced them. In some instances the form only of the organic
substance remains, and it is filled up with new stony matter.
In other instances the yellow and glittering compound of sulphur and iron, called pyrites, is found to have taken the place
of the original organic matter, and in some few instances the
unaltered parts of shells are found. The remains of the nautilus
exhibit this fact. It retains its lustre and its beauty, and
that its composition remains unaltered may be demonstrated by a
very simple experiment.

Mr. Hatchett [1765-1847; DSB 6:166][9] has shown that the
shells of living animals contain a gelatinous matter which, when
dissolved in water or an acid, renders turbid the solution of
galls. A portion of this shell has been dissolved in a little
diluted muriatic acid, merely sufficient to neutralize the
lime. When a solution of galls is mixed with it, a cloud will
be distinct, which could not have happened if the shell had
been merely lime and carbonic acid, for a solution containing
an equal weight of this fossil which--though it has the appearance of a shell--is wholly limestone, will not produce a similar effect.

The general aspect of secondary limestone is smooth and
regular and its tints are often rich and varied, but unless a
section of it is presented, it seldom exhibits much ruggedness
of outline. Its strata are generally disposed in horizontal
layers but they now and then occur inclined and in a few cases
curved or bent in the form of arches. Under these appearances,
limestone exhibits a strength and boldness of character scarcely
inferior to those of the primitive rocks, and the effects of the
mixture of the regular with the disordered strata, and the relations of the curved lines to each other, are well calculated
to awaken the strong emotions connected with the perception of
natural beauty.

Secondary schist, like shell limestone, is found in almost
all districts that contain organic remains. It varies considerably, however, both in its texture and arrangement. The hard
schist is principally composed of siliceous matter, but its
colouring substance is iron; its hue is generally dusky blue or
black. It does not effervesce with acids and is difficultly
scratched with a knife.

The soft schist is much more scaly, and less firm in its
texture, and has no character of crystallization. Its colour
varies from pale grey to deep black. The lighter specimens are
coloured with iron and the darker generally contain coaly matter.

In the hard schist very few organic substances are found,
but soft schist often contains the remains of fishes and almost

always the impressions of plants, in which carbonaceous matter is
evident. Those vegetable impressions vary in their nature in
different strata but the most common species are similar to
ferns.

Instance

The soft schist is almost always disposed in horizontal
layers. Its height is seldom considerable nor its appearance
striking. But the hard schist exhibits a much more interesting
aspect and it often occurs in irregular diffused beds, some of
which are perpendicular, as if raised from their original positions, and others of which are curved in a manner similar to
limestone.

Instance--Lumsden

The cliffs represented in this picture are nearly 300 feet
high and in some places quite perpendicular. Their masses are
large, their interstices considerable, and their outline is
broken from decomposition.

Sandstone is a secondary rock very various in its texture
and composition, but it uniformly consists of particles of sand
or gravel cemented together either by calcareous or siliceous
matter. It is seldom very firm; when calcareous, it effervesces with acids; when siliceous, it scratches glass.

Instance

When pebbles are embedded in stones of this kind, they are called
plumb [sic], pudding stones or breccia. It is one of the substances most abundant in organic remains and often contains the
impressions of entire fishes, various shells, and the parts of
land and of marine plants.

Instance

In some few instances the remains of quadrupeds even are found
in this rock. The entire skeleton of a crocodile was lately
discovered in the calcareous sandstone between Bristol and Bath.

In this work, published in 1726, there is an account of a
skeleton found in a sandstone rock, supposed to be a human
skeleton.[10] The paper is entitled "An account of the remains of
a man who had witnessed the deluge," by Jacob Scheuchzerus
[1672-1733; DSB 12:159], but the examination of the plate will,
I believe, induce every impartial enquirer to doubt if the bones
were human. To me they seem to resemble much more the remains
of a fish than of a man, or at least it must be allowed that the
antediluvians, if they had such skeletons, must have been very
different from the present inhabitants of the globe.

Sandstone rocks are much more regular in the arrangement of
their layers than any other class of secondary substances. They
are almost uniformly disposed parallel to each other, and their
aspect is generally smooth and uniform, and their fracture generally square.

Instance

Pit coal is always found in secondary countries. Its basis
is the same substance as charcoal, but it is usually impregnated
with bitumen which is the part of the coal that fuses and occasions caking and that emits the vapour which produces flame.

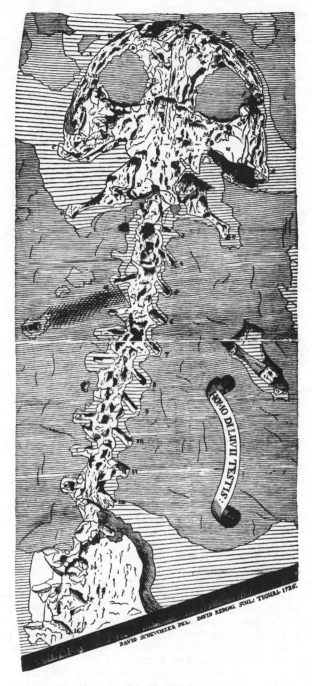

Figure 6.2. "Homo diluvii testis" appeared originally in Johann Scheuchzer's *Homo Diluvii Testis* (Tigura, 1726). Reprinted by permission of MIT Press from *Toward A History of Geology,* ed. Cecil J. Schneer (Cambridge, 1969), fig. 1, p. 192.

Coals differ very much according as they contain different proportions of fixed coaly matter and of bitumen. And they are often adulterated by earth and by the compounds of sulphur and iron, which I have already mentioned under the name of pyrites. The best coal is that which burns with a bright flame, the fumes of which do not blacken silver held over them, and which leave but a small quantity of earthy residuum.

Coals often contain vegetable impressions and sometimes the remains of animals. And in certain states it is found possessing the texture of wood which is, as will be shown in the next lecture, a strong indication of its origin. Coal generally occurs in thin layers, or seams, which seldom extend to any considerable depth. Its appearance in masses is too well known to need any elucidations, and it cannot be mistaken for any other secondary stratum.

The last great class of the secondary masses that I shall notice in this lecture is whin, or basalt. These names are applied to very different substances, but in their philosophical meaning they ought always to signify stones consisting principally of hornblende and feldspar, not so perfectly crystallized as in syenite but in other respects the same as the constituent parts of this primitive rock.

In this specimen of whin, or basalt, the black spots are hornblende; the white, the feldspar. Here the distinct crystals are large, but the gradations in size may be traced till no visible form can be perceived, as in this stone which is a piece of compact basalt from another part of the same rock.

Basalt is fusible at a white heat, is always sufficiently hard to cut glass, does not effervesce with acids, and contains more than half its weight of siliceous earth. Large masses of basalt often occur containing no organic remains, but in a few instances vegetable coal is found in it and, now and then, the impressions of fishes.[11]

Basalt is distinguished from all other rocks by the variety of its aspect and by the singularity and beauty of its arrangement. It sometimes occurs in layers between the other secondary rocks, but in this case it is usually split perpendicularly so as to form distinct parts of irregular shapes.

Instance

At other times it is arranged in columnar masses which are connected as series of regular pillars, and this is the most interesting and impressive form in which it appears.

The island of Staffa on the western coast of Scotland presents one of the most magnificent views of this kind that has ever been observed in nature. The coasts of Staffa are about three miles in circumference and all their upper parts are composed of ranges of basaltic pillars, differing in their size and height, but all in regular forms.

In Fingal's Cave in the southwest part of the island, the columns extend to the height of 50 feet; they are principally six-sided and some of them 3 feet in diameter. This picture will represent this wonderful natural excavation. It is be-

Figure 6.3. The Grand Cavern, Staffa. From Thomas Pennant,
A *Tour in Scotland, and Voyages to the Hebrides, in 1772* in
A *General Collection of the Best and Most Interesting Voyages
and Travels,* ed. John Pinkerton, 6 vols. (Philadelphia, 1811),
3: 171-569. This drawing appears on p. 305.

tween 3 and 400 feet in length and, from the surface of the
waves to the top of the cliff, is nearly 200 feet.

The pillars are for the most part six-sided but some few
occur with five sides and some with seven. Besides the great
series of columns, there are several smaller rows round the
mouth of the cave placed one above the other, and the hollows
in the lower columns are fitted to receive the projections of
the upper columns.

This is a part of one of the small series of columns which
I brought from Staffa last summer. I had the satisfaction of
passing a day upon the island. My expectations were raised to
the highest pitch with regard to Fingal's Cave by the descriptions of Sir Joseph Banks [1743-1820; DSB 1:433], Dr. Von Troil
[1746-1803], and the Honourable Mrs. Murray;[12] yet the scene
exalted the highest sentiment of admiration. With all the regularity of a work of artificial production, it combines a boldness and a variety which it is in vain to seek for in human
labours. Nothing can be more simple or more grand than the
combination by which it is formed and on this circumstance its
strength and duration depend. A few perpendicular and solid
pillars support the whole of the weighty mass above, and yet
they are so constructed as to resist the waves of the western
ocean by which they are constantly surrounded and to the whole
force of which they are constantly exposed.

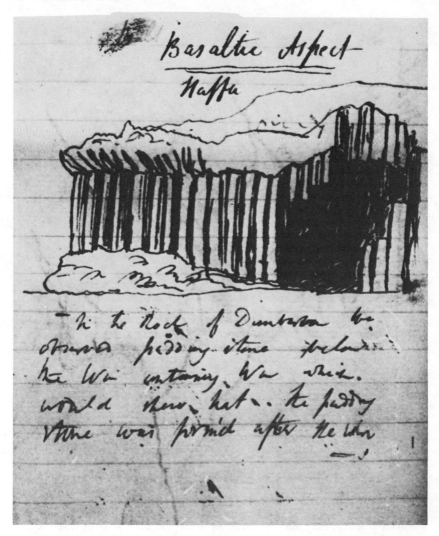

Figure 6.4.   Basaltic Aspect, Staffa.   The lines below Davy's drawing read: "In the rock of Dumbarton we observed pudding stone below the win [whin] containing win which would show that the pudding stone was form'd after the win."   Davy MSS (15e, 150) 1804.

Figure 6.5. Pleaskin. This is one of the many headlands near
the Giant's Causeway along the northern coast of Ireland where
the columnar basalt of which Davy writes occurs. He did not
visit these formations until the summer of 1805. William Hamilton
Drummond, *The Giant's Causeway, A Poem* (Belfast, 1811), p. 33.

Figure 6.6. Davy's drawing of Pleskin [Pleaskin]. Davy MSS
(15g, 97) 1805.

Figure 6.7. Eastern View of the Giant's Causeway. From William Hamilton Drummond, *The Giant's Causeway, A Poem* (Belfast, 1811), p. 1.

Figure 6.8. Win Dikes, Causeway. Davy MSS (15g, 95) 1805.

The Giants' Causeway on the coast of Antrim in Ireland is a basaltic arrangement almost as remarkable as that of Staffa. Many series of great columns occur one above the other, and they are for the most part articulated or composed of points.

This is a print of the Giants' Causeway. It exhibits the great extent and general arrangement of the basaltic rocks which in this part of the island form the principal cliffs of the coast.

The basalt of Antrim is similar in its texture and composition to that of Staffa. But in the basalt of Eigg, an island to the north of Staffa, pitchstone occurs in the rock mixed with the feldspar and hornblende, and the columns in general have fewer sides. These columns, however, in the high basaltic mountains of the island called Eigg, are piled together with astonishing grandeur, and they form a precipice of several hundred feet in height.

Instance
The basaltic rocks alternate with other rocks in the general scene. The whole picturesque effect is consequently increased and that delightful emotion of surprise is awakened in the mind, which always results from a perception of apparent art amidst the rude forms in nature, or of nature amidst the refinements of art.

The relations of the positions of the different secondary rocks that have been described are almost as diversified as those of primitive rocks and vary in particular districts. Limestone is the secondary rock generally discovered at the greatest depth and schist or sandstone are usually found placed immediately upon it. Coal is sometimes discovered upon limestone but hardly ever under it; the strata commonly above coal are sandstone and loose schist, and it sometimes occurs beneath basalt. A red sandstone of this kind or a schist of this kind containing vegetable impressions almost certainly indicates the substance.

Instance
In Derbyshire, where the surface has been penetrated to more than 1,500 feet, seven or eight regular strata only have been found. Of these the first is sandstone; the second, brown limestone; the third, schist; the fourth, coal; the fifth, sandstone; the sixth, black limestone; and the seventh, grey shell limestone. The basalt in Derbyshire, or the "toadstone" as it is vulgarly called by the miners, forms no regular stratum but occurs in masses or veins amidst the strata.

Instance
In Durham and Northhumberland and through the great extent of Alston Moor in Cumberland, the secondary strata that occur are heaped one on the other in parallel layers and are more numerous than in any other district. In Teesdale, within 1,200 feet in depth, 121 strata occur, which are limestone, sandstone, soft schist, hard schist, basalt, and coal, various in thickness and occurring in an irregular order.

Instance

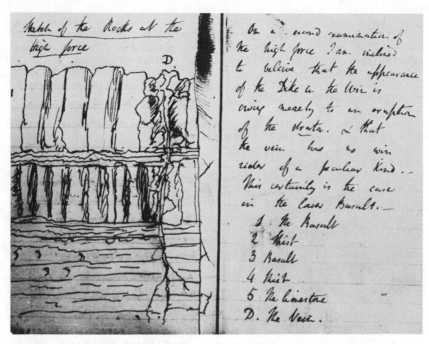

Figure 6.9. Sketch of the Rocks at the High Force. "On a
second examination of the high force I am inclined to believe
that the appearance of the dike in the win [whin] is owing
merely to the eruption of the strata, and that the vein has no
win rider of a peculiar kind. This certainly is the case in
the lower basalt. 1. The Basalt 2. Shist 3. Basalt
4. Shist 5. The Limestone D. The Vein." Davy MSS (15e,
108-9) 1804.

The extent of the strata may be judged of from the appearance in
the painting of the fall of the Tees, which has been already
shown. The rocks which are represented here form a part of the
general layers of the country. They are not more than 100 feet
high, taking the bed of the river as a level, and yet they contain no less than seven distinct layers of rock.

Instance

The chalk strata are generally distinct and insulated and
seldom alternate with any of the other rocks.

The strata of the secondary rocks in no case blend or pass
into each other by mixture or transition, but they are often
wholly separated by thin layers of soft clay or sand. This is
particularly the case in Derbyshire.

Instance

When schist and limestone meet, they are merely loosely attached
to each other. But sandstone formed upon limestone often adheres
to it with considerable tenacity, though there is never any mixture of their parts.

Figure 6.10. Fingal's Cave on the Isle of Staffa. From L.A.
Necker de Saussure, *Voyages en Ecosse et aux Iles Hébrides*,
3 vols. (Geneva, 1821), 2: 302.

When basalt is found upon sandstone or upon pudding stone,
which is very often the case, they are always firmly united.

Instance

In Staffa the basaltic pillars rest upon a species of imperfect breccia consisting of sand pebbles and fragments of basalt united by a siliceous cement containing much iron. And
the bases of the columns are fixed and, as it were, cemented
into the irregular mass beneath. The junction of basalt with
breccia, or as it has been sometimes called, "basaltic tufa,"
in Staffa is well exhibited in this picture which, as well as
the other pictures of basaltic rocks, were designed by Mr.
Webster [1773-1844; DSB 14:210], and it perfectly represents
the union of the two classes of rocks and gives an excellent
idea of the general form of the island.[13]

Instance

The influence of the great arrangement of the secondary
mountains in the general economy of nature is in some respects
similar to that of the primitive mountains. They increase the
extent of surface and they are subservient to the equalization
of the temperature of the climate in which they exist. More
easy of decomposition than the primitive rocks, they more
readily become the abode of vegetables. They supply the waste
of land taking place in the valleys and plains. And from the
mixture of their parts brought into a state of minute division
by the agency of moisture and the atmosphere, a soil is provided the best adapted for the support and nourishment of
plants.

All the parts of the softer secondary strata are in a certain degree permeable to water, and in consequence of their
power of retaining this fluid so as to suffer it gradually to
percolate through them, they remain saturated with it in all
seasons and are the principal sources of those streams which
flow in the heats of summer and the driest weather, and which
are absolutely essential to the existence of vegetables and
animals.

Lecture Seven

Whenever any circumstances occur which mark distinct periods or a succession of events in nature, they uniformly supply means of investigation and enable us to reason concerning the relations of their causes. In considering the secondary strata filled with organic remains and containing the vestiges of many beings analogous to living generations, it is scarcely possible to avoid speculation. Even the most sober and cool understanding is almost irresistibly led to form ideas concerning the state of things existing at their origin and concerning the powerful material agents subservient to those great natural events, the occurrences of which are demonstrated by their arrangement and wonderful appearances.

The subject is one of high interest, but it ought to be pursued with considerable caution. Men possessing very brilliant talents have differed essentially upon it. Hypotheses have been multiplied with regard to it, and much ingenuity and labour have been employed upon the discussion.

In accounting for the present order of things, we are enabled constantly to appeal to the senses, to combine observation and experiment. But in attempting to explain the past changes in nature, if we are in possession of no distinct history, analogy is our only guide. Inferences must be made from the known concerning the unknown facts. And the mind, if unable to obtain certainty, must rest satisfied with strong probability.

None of the facts concerning the formation of the secondary rocks can be immediately known. All our reasonings upon them must be inductive. They must be founded upon the knowledge of the

properties of the substances that compose the strata and of the
agents by which these substances are capable of being modified.
And where causes can be plausibly developed, it must be from the
examination of existing operations that produce appearances similar to those which are the objects of enquiry.

Amongst the secondary rocks, limestone is that which occu-
pies the most important place, and it is the stratified substance
perhaps the most interesting in its relation to general theory.
All shell limestones when minutely examined exhibit a texture
which proves that their constituent parts, at the time they were
deposited upon the organic remains, were in a state of liquidity
or solution.

This inference, it is not possible to doubt. The interstices of the shells in marble are almost always filled with crystals. Crystals likewise abound in various parts of most of
these stones. As their formation must have been subsequent to
that of the organic bodies they include, it implies a previous
state of chemical division. This is a principal fact admitted in
all the theories of geology, but explained in different manners.

The only two agents which act upon a great scale in nature
in producing fluidity or solution are fire and water. These,
being the only powers which could have operated in the production
of the secondary strata, at least known to us, are the only
powers concerning the agencies of which we have a right to conjecture.

It has long been known that limestone, carbonate of lime, is
soluble to a small extent in water. Almost all our common pump
and river waters contain it, and it is more soluble in water impregnated with carbonic acid, or fixed air, than in pure water.
It has likewise been shown by some late experiments that this
substance is capable of being fused by heat.

Under common circumstances when limestone is ignited in our
fires, the carbonic acid, which forms almost half of its weight,
flies off and the pure lime [remaining] is wholly infusible. But
when limestone is heated strongly in closed vessels from which
the elastic fluid cannot escape, the compound is readily softened
and becomes fluid at a white heat.

This is a beautiful discovery of Sir James Hall. I have had
the pleasure of examining his specimens and of witnessing his
method, and I have seen pieces of compact stone even containing
crystals which have been formed by heat under pressure from
coarsely powdered chalk.[1]

As both the great agents of nature are capable of rendering
fluid the constituent parts of carbonate of lime, we must be
guided by appearances in endeavouring to decide which was most
probably concerned in the production of the secondary limestone.

The shells contained in limestone preserve their perfect
forms and some have their texture unaltered. Now it is difficult
to conceive that this could have happened if they had been surrounded by melted matter, for they must themselves have been
brought into fusion and, consequently, ought to be found of the
same texture and compactness as the stone in which they are im-

bedded, or at least they ought to have been in some measure altered in their form or compaction.

The greater number of shell limestones are disposed in alternate layers, and these have every appearance of having been
produced by a deposition from water. The effect is very little
compatible with the idea of igneous fusion of the materials of
the strata.

The strongest argument that has been produced for the agency
of fire is derived from the fact of the curvature of some of the
calcareous strata. This, it has been said, almost proves that
they were in a soft state and that they were elevated at the time
of their formation by some force acting from beneath.

Instance

These are the sketches of the arched strata; you saw them in
the former lecture. To me they seem unconnected with any effect
of heat, and they may easily be accounted for by supposing an
arched form in the surface upon which the first layer was deposited.

There are analogous instances which occur every day. This
is a piece of carbonate of lime, having all the properties of
limestone, which was formed in the boiler of a steam engine
supplied with the water of the Thames. When magnified it appears
wholly composed of curved layers, and this sketch represents a
section of it enlarged about five hundred times.

The same general conclusions may be applied with regard to
calcareous sandstone, which consists of particles that had not
been in solution cemented by carbonate of lime which had been
in solution, so as to assume a crystallized form.

On the subject of siliceous sandstone a new order of facts
must be brought forward. The constituent parts of this substance
are infusible by any heat that has been as yet applied in our artificial processes. But siliceous matter is soluble under certain circumstances: in water, when heated or containing alkali.
Thus the boiling water which spouts from the Geysir in Iceland
was found by Dr. Black [to be] impregnated with siliceous earth.[2]
And siliceous earth appears to be capable of being dissolved by
the fluids of plants, for I have detected it deposited in considerable quantities in the interior of various grasses and in rat-
tan. And Mr. Macie [1765-1829; DSB 12:494] found large nodules
of siliceous concretions within the points of the bamboo.[3]

That siliceous sandstone cannot owe its compactness or its
crystallized texture to the agency of fire, even allowing it capable of being fused, is, I think, evident from many appearances.
Thus there are often found embedded in siliceous sandstone, fragments of very fusible rocks not apparently altered or rounded in
their forms. This I have seen in several instances. I particularly recollect one in a quarry near Edinburgh, in which pieces
of whin of very different shapes occurred in the rock, and whin
is fusible at a heat which would scarcely affect the sandstone.

As siliceous sandstone is found both above and below shell
limestone, it is scarcely possible to doubt but that some of the
same causes must have operated in the production of both these

Figure 7.1. The Rocks at Lumsden Burn. The strongly folded strata near Lumsden Burn, like the unconformity at nearby Siccar Point, had caught James Hutton's attention some years before Davy gave these lectures. Hutton published a drawing by Sir James Hall very similar to this one by Necker de Saussure, among the plates at the end of the first volume of his *Theory of the Earth* (Edinburgh, 1795). Necker de Saussure, like Davy a few years before him, was following almost literally in Hutton's footsteps, for he too was guided by John Playfair and Sir James Hall to these sites. From L. A. Necker de Saussure, *Voyages en Ecosse et aux Iles Hébrides,* 3 vols. (Geneva, 1821), 1: 280.

Figure 7.2. Arch in the Strata near Lumsden. Davy MSS (15e, 123) 1804.

classes of stones. And consequently the only probable agent to
which we can refer the solution of the siliceous matter is water,
either heated or under some peculiar circumstances of combination.

The different species of schistose rocks present no crystallized texture; some of them are capable of being fused and most
of them are hardened by fire.

Instance

There is consequently very little reason to suppose that this
agent was concerned in their production, and their origin may,
with a much higher degree of probability, be referred to a deposition of minute solid particles of argillaceous and siliceous
matter which had been mechanically suspended in water and of
which the present adhesion is owing to bituminous matter, or
oxide of iron.

Chalk, which consists of particles of carbonate of lime
merely in a state of mechanical adhesion, must, as there is every
reason to believe, like schist, have been produced by the deposition of solid matter finely divided and suspended in a fluid.
With regard to the strata of flint which generally alternate with
chalk, it is very difficult to assign any satisfactory cause for
their formation. That they could not have been produced by fusion is evident, for if their constituent parts are at all capable
of being rendered liquid by heat, it can only be by a degree much
more intense than we are able to produce. And such a heat acting
under pressure must necessarily have likewise fused the calcareous matter surrounding them. It very often happens that perfect
impressions of the *Echinus* are found filled with flint, whilst
their exteriors are crystallized carbonate of lime.

Instance

Now it is scarcely possible to conceive that this could have
taken place unless the siliceous matter had been carried through
the orifices of the shell and deposited in the cavity by a fluid
and gradually acting menstruum.

On the subject of the origin of coal there have been various
opinions. But the vegetable and animal impressions that it contains, as I mentioned in the last lecture, ought to lead us without much minute discussion to the theory of its production.
Where wood has been buried in the earth in late times, from the
action of water and of air it always undergoes a change which
brings it to a state similar to that of pit coal. At Bovey in
Devonshire a whole forest is found buried under different strata
of clay and gravel, and the trees, though carbonaceous and black,
have their ligneous form and contain a substance apparently intermediate between bitumen and resin, as has been shown by Mr.
Hatchett.[4]

A similar species of coal is found in Sussex and in Iceland,
where it is named "surtarbrand," and the pieces appear flattened
from the pressure of the superincumbent mass during the time of
the conversion of the wood into coaly matter.

Vegetable and animal substances, however, may likewise be
converted into a body resembling pit coal by heat under pres-

sure. Sir James Hall has shown this, making use of the same
means as those employed in his experiments upon limestone. He
has proved that, by a strong heat applied to different animal
matters and different species of wood in closed vessels, a substance is produced, black, and glossy, and having properties
different from those of common charcoal, as it burns with a
dense flame and gives out volatile matter.

Sir James has adduced his experiments as proof of the Huttonian theory of the igneous origin of pit coal. His arguments
are in the highest degree ingenious and, if the insulated fact
only were to be considered, very forcible and convincing. But
when the usual strata above and below coal are examined, it is
scarcely possible for the impartial observer to suppose that they
had been subjected to heat. They are principally loose sandstone
and loose schist, bearing every mark of having been deposited
from a state of mechanical suspension in water.

Instance

The agency of water and the natural process of decomposition to
which all organized matter is liable thus seems fully adequate to
explain the appearances of the coal strata.

Vegetable matter in its first state of fermentation appears
as peat. In another and more advanced state it is the same substance as Bovey coal, and when the bitumen is completely formed,
it is the true fossil coal in which the ligneous texture is
sometimes evident and sometimes destroyed.

In a quarry near Glasgow, vegetable matter is found in almost all its different states of conversion. Near the surface it
is but little decomposed and, here, it occurs in a very imperfect
sandstone, apparently of late formation. In a lower stratum,
about 10 feet below, it has the appearance of wood slightly
charred. And in the compact and white sandstone which is used
for building, it occurs apparently converted into bituminous coal.

Instance

The last subject of particular theory which it will be
necessary to discuss is perhaps the most difficult and the most
connected with the operation of unknown powers. I mean the formation of basaltic rocks. For a long while, all basalt was considered as a volcanic production. But since it has been found
alternating with other rocks deposited in regular layers, this
generalization has been given up by a number of philosophers and
a more limited theory adopted.

Basalt was shown, in the course of the last lecture, to be a
crystallized substance consisting of hornblende and feldspar.
Its elements are not soluble in common water, but they are capable of being rendered fluid by heat, though after fusion they
have never yet been made to assume their original form. It is
not possible therefore to account for the origin of the rock satisfactorily from any certainly known facts, and we must be contented with presumptive evidences from the particular appearances
which it presents and the circumstances under which it is found.

All the basaltic rocks of great extent as yet examined have
been discovered upon common secondary rocks containing organic

remains. And these layers of basalt, extending over many square
miles, often occur between strata of schist and sandstone, of the
common kind and having nothing peculiar in their aspect.

The same basaltic strata which occur in Teesdale, and which
are represented in this sketch, extend to the north into Alston
Moor. They have been traced more than 30 miles, and they are uniformly found between the same schistose strata. The schist below them and incumbent upon them contain organic remains and is
exceedingly soft, though when acted upon by a heat much below
that necessary for fusing the basalt, it becomes very hard.

Instance
And at the white heat, it is itself capable of being rendered
fluid.

Instance
So that it is impossible to suppose that this basaltic stratum
was ever in igneous fusion.

The same reasoning may be applied in accounting for the arrangement in the other layers of basaltic rocks found in the same
district. I have examined three of the lower strata and, to me,
they have uniformly appeared to exhibit analogous facts.

In the Western Islands of Scotland and on the coast of
Antrim, all the great and majestic arrangements of basalt are uniformly incumbent upon rocks which exhibit no marks of the agency
of fire. At the Giant's Causeway the basalt is said to alternate
with sandstone, and at Staffa, as appears from the sketch and as
I mentioned in the last lecture, it is placed upon a breccia
which contains, immediately below the pillars, fragments of a
stone equally fusible with basalt.

Instance
The basaltic rocks in the neighborhood of Edinburgh are almost all fixed upon sandstone, schist, or limestone, but their
arrangement is much more disturbed than in most other cases. In
the basaltic mountains called Arthur's Seat, several layers of
irregularly disposed columnar basalt are formed alternating with
sandstone. And at the point of junction, the sandstone is of a
different texture, considerably harder, and of a red colour.
This fact was adduced by Dr. Hutton as a demonstration of the igneous origin of basalt. He supposed that the materials of it,
being projected in a fluid state amongst the strata of sandstone,
had hardened them at the point of contact and given them a new
texture.

I have paid particular attention to this appearance. I exhibited a specimen of the junction of the basalt and sandstone in
the last lecture. This is a piece of the general layer of sandstone in its softest state. I have exposed a part of this layer
to a heat sufficient to fuse basalt, but I do not find that it
indurates in any considerable degree. And in examining the hard
mass in contact with the basalt, and in comparing it with the
soft mass below it, I find that their difference does not depend
upon a different consolidation of the same matter, but upon a
real difference of [composition], for the hard substance contains
much more siliceous matter than the soft one.[5]

Some of the pillars of the great basaltic arrangements often occur in a bent form, and curved, and this has been brought forward as an argument for their having been fluid. But the same explanation may be given of this phenomenon as of the other phenomena of the curvature of strata.

The columnar forms generally speaking indeed offer no evidences in favour of the igneous fusion. It is said that some of the lavas of Etna, where they have flowed into the sea, are found in the state of pillars but these are very rare instances.

Similar effects take place much more commonly during the drying of masses which have been consolidated by the agency of water. Thus the common form of starch is produced at the time of its drying, and if pure clay be made into a paste with water and spread upon a large surface, at the time of the evaporation of the moisture it always splits into pieces, many of which have regular forms.

All the facts seem to render it probable that the basaltic rocks usually found in alternation with the common secondary rocks cannot be of igneous origin. But the proposition of the impossibility of the formation of this substance by fire must not be advanced. In many secondary and primitive countries, veins or dikes of basalt are found which have many characters similar to those of lava and which often appear to have produced similar effects.

At Cockfield in Durham the coal strata are intersected by a vein of basalt, and where this substance comes in contact with the inflammable matter, its effects are precisely similar to those which would be produced by fire. The bitumen of the coal is driven off and the coal appears as coke. In Northumberland and in Derbyshire, there are analogous instances.

Instance

I have had but few opportunities of examining the kind of basalt which fills these dikes, but in the specimens that I have seen, it has appeared of a much finer grain than the common basalt and approaching nearer in its nature to lava.

Instance

I have already mentioned that basalt may be fused and after fusion may be made to crystallize by slow cooling, though its constituent parts do not separate into the distinct forms of hornblende and feldspar. Whether the crystals are ever perfectly distinct in the basaltic dikes, I am not able to decide. Sir James Hall, by preserving basalt in a state of imperfect fluidity for a considerable time, at last procured a body exceedingly analogous in its general appearance to the original stone. His experiments were repeated upon a much larger mass of the same substance by Mr. Gregory Watt.[6] There is in this drawer a series of the specimens produced in different states of gradation from a mere opaque glass to an assemblage of uniform crystals.

Mr. Watt's paper may be found in the last volume of the *Philosophical Transactions,* and it is filled with curious reasoning and with useful information. It was the first and the last production of an early matured genius who had devoted him-

self to geology and who had given the strongest promise of pro-
moting its most important objects.

Mr. Watt had examined a number of the most perfectly formed
volcanic specimens with great attention. He had accurately compared them with the stratified basalt, and he was fully of the
opinion that substances very nearly alike might be formed both by
aqueous deposition and by igneous fusion. He had himself no particular hypotheses; he had opposed the different facts to each
other with great fairness and caution, and he was preparing a
series of experiments to elucidate the general history. Unfortunately, he did not live to execute his designs and those only
who were intimately acquainted with the extent of his knowledge,
the quickness of his perception, and the accuracy of his judgement, those only, can estimate the serious loss that science has
sustained in his recent death.

I have now stated, and I hope without any prejudice of system, those opinions belonging to the different theories which
offer the most plausible explanations of the causes that appear to
have operated in producing and consolidating the particular secondary strata.

Concerning their general arrangement and the revolutions in
nature connected with it, I have as yet offered no supposition,
but this part of the subject necessarily follows in the order
of discussion. All the appearances, all the different facts,
show in the most decided manner that the various strata filled
with organic remains were covered by the sea at the time of their
production. This is the ancient theory which will not be dis-
puted, though many modifications of it may be formed, and various
arguments must be stated before it can be decided whether the
strata were formed by slow operations or by rapid changes, whether
they were the works of days or of ages, and whether the present
land was anciently the bed of the ocean or whether the secondary
rocks were formed upon it by a great influx of waters.

Mr. Deluc, a part of whose theory I noticed in a former lecture, in his last publication, the letters to Blumenbach,[7] attempts to show that the secondary rocks were all produced long before the creation of man. He conceives that the original ocean
held in solution all the matters that compose [the secondary
rocks] and that these matters were deposited and consolidated
at the time that the waters were peopled with marine plants and
animals. Reasoning upon the successive depositions of strata, he
supposes that an immense length of time must have been essential
to their formation, and he believes that the six days in Genesis
mean periods almost indefinitely great.

It is with regret that I am again obliged to express an opinion very different from that of this venerable philosopher. But
in appealing to the most simple and obvious analogies only, his
hypothesis seems incapable of being reconciled to the appearances.
Supposing that the ocean had held in solution the different substances which form the secondary strata, in Mr. Deluc's theory
there is no cause for their precipitation, no new agent is supposed to be introduced; and as the same bodies uniformly preserve

the same attraction for each other, the series of events that he has assumed is contradictory to the established chemical laws. Besides, supposing a fluid saturated with siliceous matter, with the other earths, and with the metals, it is scarcely possible to conceive that it was fitted for the purpose of life or that it could have been the abode of marine animals at the time the great changes of composition which Mr. Deluc imagines were taking place in it.

The existence of coal strata of vegetable and animal origin alternating with the other secondary rocks seems almost decidedly to show that their formation took place upon a surface which had before been habitable land. The idea of Buffon, that the carbonaceous matter might have been washed down by rivers and buried in the bottom of the sea, will not coincide with certain simple facts. In coal, and in the strata above and below coal, are found perfect impressions of the thinnest and most delicate vegetable leaves, and such substances so easy of decomposition, if derived from the land by rivers, must necessarily have been destroyed by the action of air and water long before they could have reached the depth of the ocean.

There are no distinct appearances which demonstrate a very slow and gradual deposition of the secondary strata, but there are many which show that, at least in some cases, the effect of their formation must have been sudden and rapid. Fishes, the organized matter of which so easily consumes, have left their impressions embedded in the hardest rocks and these impressions often exhibit the perfect and unaltered form. Other marine animals of a still more delicate organization likewise exhibit in the vestiges of their remains an undecomposed structure.
Instance
The various appearances would seem to indicate that these different beings had been suddenly destroyed and suddenly enclosed in the strata that contain them.

The various remains of the land animals found in sandstone, alternating with shell limestone, offer another indication of the waters having covered the place of their abode. And that these waters flowed in upon a great extent of country and rushed over the surface with considerable force seems evident from the transportation of the remains of the animals which are now only found within the tropics, toward the poles. The bones of the rhinoceros and of the elephant have been found in Siberia. The tusks of the elephant have often been discovered in this country. I mentioned in the last lecture the bones of the crocodile found at Bath. I have since been so fortunate as to procure a drawing of them, made by Mr. Rickharts, from the kindness of Sir George Paul.[8]

Instance
Astronomical deductions seem to show that these appearances cannot result from a change of climate arising from a change of the inclination of the axis of the earth. And when all the evidences are examined and all the various oppositions and arguments of facts discussed, there seems to be no period to which the pro-

duction and arrangement of the secondary rocks can be so well
referred as to that of the great inundation of the waters upon
the land, recorded both in sacred and profane history, of which
so many testimonies are preserved in nature, and of which so many
traditions have been brought down from the elder nations and from
the most remote times.

We perceive the effects of this great catastrophe, but the
immediate natural cause of it can never be distinctly developed.
The hypotheses of Leibniz, extended by Whiston, that it was produced by the attraction of a comet upon the waters of the ocean,
is perhaps the most plausible that has been advanced and, when
taken with limitation, the most adequate to the explanation of
the phenomena. Supposing with Leibniz that a large comet ap-
proached near the earth, having a power of attraction sufficient
to raise the tides to the utmost height of our mountains, and
supposing likewise that it was on its return from the sun and
capable of communicating heat to the waters, its influence, there
is great reason to believe, would be fully adequate to produce
the various effects that have been observed and to occasion those
great and diversified changes exhibited by the secondary surface
of the globe.

Assuming an elevation of temperature of only 100 degrees
where the ocean was elevated into tides, which in itself would
not be destructive to life, there is great reason to suppose
that the solvent agency of the water would be sufficiently increased to enable it to combine with the different earths,
though the other parts of its surface, from its bad conducting
power, might remain comparatively cold. And from the mixture
of the heated and the cold fluid, different depositions would
occur and very different strata be formed.

The water of the ocean, in rushing over the different parts
of the land, would necessarily carry with them their living inhabitants and many of the substances forming their beds. Where
forests were overwhelmed, layers of coal would be produced;
where calcareous soils were torn up, strata of chalk would be
formed; and the sand from the shores and from the depths of the
sea deposited at the same time with dissolved earthy matter would
produce the varied cemented stones and breccias.

The general effect of the action of the sea would be increased by the rapid evaporation at the point of attraction and
from the depositions of torrents of water in the more remote
parts. In the great series of changes, rocks that had been covered by the sea would be laid bare, many high lands would necessarily be broken down and levelled into plains, the forms of
mountains would be changed and the productions of one country
carried into another by the rapid impulse of the waves.

It would be easy to offer still more minute elucidations of
the opinions of Leibniz, but to multiply imaginary instances is
perhaps to indulge too much in the spirit of speculation. Whatever may be the cause assigned for this great event, it must be
considered as one independent of the common order of nature, for
the usual change of the sea and the agencies of the land nothing

analogous occurs, and it must be referred to some power operating
in a novel manner in our systems. As yet we are not perfectly
able to explain the existing economy of things and our theories
of past changes are remote. Placed, as it were, mere atoms upon
a point of space, we perceive only a few objects and a few of
their relations, and to these we are obliged to refer in all our
reasonings. Whilst in the universal series of occurrences, influences may have acted and may still be acting of which we have
no conception.

Yet nevertheless it is proper that we should reason from the
present concerning the past, if we reason with a calm understanding. The strength and the correctness of the imagination can
only be preserved by exercise, and suppositions, when they are
made only the amusements of the imagination, are rather useful
than injurious, for they increase the activity of the mind; they
accustom it to rapid combinations. And they are only dangerous
when they are insulated from facts, when they are mere distempered dreams, or when they are pertinaciously adhered to and
opposed to the convictions of truth.

The human mind, deriving all its ideas from the senses when
in a state of healthy exertion, sooner or later uniformly refers to facts. And when hypotheses are used merely as instruments for comparing facts and for ascertaining their minute relations, they promote in the highest degree the efforts of in-
ventive genius and tend to impress on the understanding the true
and unperverted images of nature.

Lecture Eight

The progress of civilization is immediately connected with the application of the metals, and it is not therefore surprising that their discovery should be mentioned in the first historical and authentic documents of antiquity. In the sacred writings of Moses four metals are mentioned, gold, silver, brass and iron, and the descriptions in Exodus prove that the art of working the metals had obtained a high degree of perfection even in the earliest era of the progress of the power of the Israelites.

The metallurgical arts appear to have been brought into Greece from the East, and the veneration in which the introduction of them was held is evident from all testimonies. In the writings of the poets, the discoverers of the metals are raised to the rank of deities and the real glory and ability of their inventions obscured by accounts which it is equally impossible to believe or interpret. In the *Iliad* and the *Odyssey*, the precious metals are often mentioned, and iron, tin, brass, and lead are likewise stated as being in common use. Homer gives a particular account of the methods of fusing, casting, polishing, and tempering the metals but makes no mention of their ores, nor of the positions in which they are found, nor of the means by which they were procured.

The first Greek author who speaks of the knowledge of metallic veins is, I believe, certainly Aristotle. But he mentions them only vaguely in his account, *Of Wonderful Reports*, to which he often appears himself to give no credit.

The mines of silver in the mountains of Spain, says the philosopher, are recorded as having been discovered by some shepherds

who, having set fire to a certain woods for the purpose of clear-
ing the ground, found, after the combusion ceased, a large fissure in the surface containing an ore of which the upper part
appeared as pure silver. And an earthquake happening soon after
laid bare the whole of the vein which was then worked to great
advantage.

Similar accounts are given by various Roman authors who
all make the discovery of mines the result of accident.
Lucretius, in speaking of the effects of the metals on the progress of cultivation and refinements, expressly adopts this
opinion. I have made a translation of the passage.

> Here were the glittering veins of treasure metal found
> Where active fire had scorched the solid ground,
> And to the eye disclosed the precious store,
> The fluid metals and the native ore.
> Such was the simple and the happy cause
> Whence the bold miners' dangerous art arose.
> From one event and one ingenious thought
> Soon was the spirit of invention caught,
> And all the treasures different riches produce
> By powerful means applied to human use.

The later Roman writers, though they often mention the various works in metals, seem to have been wholly unacquainted with
the common processes of mining. There is indeed every reason to
believe that the various mines which supplied Rome were either
worked by slaves or by the natives of the countries in which
they were found, and the inhabitants of the imperial city, provided they obtained a little iron and brass to supply the purposes of war and an abundance of gold and silver to provide for
the purposes of luxury, probably cared very little concerning
the manner in which they were procured or the situation in which
they were formed.

Diodorus the Sicilian, who wrote in the time of Julius
Caesar, in describing the productions of different countries has
mentioned the metals as generally raised and worked by the natives. He particularly describes the method of raising tin in
Britain.

These men, says he, alluding to the British miners, manufacture tin by working the grounds that produce it with great
art, for though the land is rocky yet it has soft veins of earth
running through it in which the treasure is found. The miners
extract it, melt it, purify it, and shaping it by molds into a
cubical figure, send it to the coast for exportation.

Strabo describes different species of brass as found in
Phrygia, and Pliny gives an account of the localities of many of
the metals, but he seems to have made no immediate observations
himself upon the subject and appears often to have followed very
vague authority. Thus he pretends to doubt whether any tin is
found in Britain and says that it is brought from Portugal and

Spain, where in late times not a single atom of the substance has
been found.

That the first iron mines worked in Europe were in the island of Crete we learn on the authority of Hesiod. Copper in the
earliest ages was raised in Cyprus, and from time immemorial, the
west of Britain, called by the Greeks "Cassiterides"--the island
of tin--certainly supplied Europe and Asia Minor with that useful
metal.

There seems no reason to believe that the Malacca tin was at
all known to the ancients. Aristotle indeed speaks of a white
metal brought from the East, of which the kings of Persia had
drinking cups made, but he states that it was harder and more
valuable than silver; and on the same page he calls tin the Cel-
tic metal, "κελτικον." Till the thirteenth century, tin in Europe
was exclusively raised in Cornwall and Devonshire, but in the
time of Richard, king of the Romans and earl of Cornwall, some
miners, disgusted with certain changes made by the earl in the
stannary laws, left their native country and passed into Saxony
where they discovered tin ore and taught the Germans the arts
of raising it and of converting it into pure metal.

The lead mines in Derbyshire were probably known as early as
the conquest of Britain by the Romans. But the copper mines in
Cornwall and Wales were discovered and worked only in late times.
Copper had been constantly raised with tin in the Cornish mines,
yet it was always neglected and thrown away as a base and useless
ore till the beginning of the last century when it was first
melted under the direction of a company established at Bristol.
Since that time the annual produce of the copper mines has increased so much as to exceed that of the tin mines and lately
to such an amount that it has been three or four times as great,
the produce of copper in the last year being nearly a million
sterling and that of tin only about 250,000.

We know very little of the early history of the indicators
by which the miners were directed in the search for veins.
George Agricola, whose work was published in 1561, mentions the
mineral impregnations of water issuing out of mountains as having
often led to the discovery of metallic ores. In this account and
in his general details he follows philosophical views. But in
his history of the use of the divining rod he seems to have been
led away by the popular superstitions of his age. A forked hazel
rod, says the author, when grasped in such a manner that a fork
is in each hand and held parallel to the ground, will be strongly
attracted by the metallic veins in the neighborhood, and if properly used will afford certain indications of them.

The idea of the divining rod is mentioned with great respect by most of the ancient writers on mining, and Alonso Barba
[1569-1640?; DSB 1:448], in his account of the Peruvian mines,
makes a great merit of having improved it by using two rods, one
placed crossways on the other instead of a single fork.[1]

Pryce [1725?-90], in his mineralogy of Cornwall,[2] published
as late as 1788, takes the effects of the rod for granted and en-

deavours to account for them by supposing that certain metallic
steams constantly arise from veins, though he does not make it
very clear why such steams should have a particular affection for
forked rods of hazel and for no other substance.

Mr. Cookworthy[3] of Plymouth was, I believe, the last person
of any credit who professed to be able to discover metals by the
divining hazel, and he was a very worthy and useful man and certainly deceived himself without designing to deceive others.
There are some mines on record which he is said first to have
made known by his skill, but they are all in places where veins
had been cut before, and he allowed himself that he had often
failed in his process and had often seen others fail. He had
always found the rod succeeded best in the hands of ignorant
people, or children, who would not venture to doubt of its in-
fallibility.

The want of efficacy of the divining rod cannot be better
demonstrated than by its being wholly out of use in every part of
the world. The observation of Christian of Saxony upon it is exceedingly applicable:

> An adept sent to that prince stating that he had a rod by
> which he was able to discover veins of gold and offering
> to come to his court. Christian desired him to stay
> away, saying if you are in possession of such a method
> you have no need of me. If you are not in possession of
> it I have no need of you.

The idea of the Cornish miners that metallic veins are sometimes indicated by the appearance of fires in the night over the
places in which they are found, has been generally considered as
a mere delusion. But that such a phenomenon may occur seems not
at all unlikely, for as the metallic ores are for the most part
perfect conductors of electricity, slow discharges from electrified clouds must often take place upon them and from such a cause
the effect would necessarily be produced.

It would be improper to dwell longer upon the vague and in-
determinate circumstances which have influenced the popular opinion with regard to the discovery of ore in mineral countries.
The only real indications of metallic substances valuable to the
geologist and capable of being relied on are those founded upon a
knowledge of the directions of different strata, and of the substances usually found in them, and of their relations to each
other. This part of the subject I shall now proceed to discuss
and to elucidate as far as it is in my power.

It has been mentioned in preceding lectures that the masses
of the primitive and the secondary strata are in no cases uniform
or unbroken. They all contain chasms or fissures which are sometimes empty and sometimes filled with substances different in
their appearance and composition from the rock in which they lie.
It is in this last case they are called veins.

                         Instance
So that in every vein there is to be observed the horizontal dir-

ection, the destination, the thickness, the depth, and the ramifications.

The materials found in veins are exceedingly various; crystallized stones almost constantly occur, and the veins are
scarcely ever wholly filled with metallic substances. When rich
ores are found in veins, they are usually imbedded in some peculiar stony matter different from the mass of the rock.

Instance

The same fissure often extends for a great number of miles
and intersects various rocks, but in the different rocks it is
filled with very different substances. There is a very striking
instance of the difference of the composition of a vein in its
course through different rocks at St. Michael's Mount.

Instance

Figure 8.1. St. Michael's Mount. From an engraving of 1802.

The granite and schist join on the east side of the rock and
there is a vein which runs through them both east and west, but
in the schist the vein is wholly filled with quartz and, in the
granite, it contains mica and schorl rock with a little tin ore.

Similar circumstances occur perhaps more distinctly in the
rocks of parallel stratification. The whole of the strata of a
country are often cut through by the same vein having a few
different ramifications. This represents a section of parts of
the strata in Alston Moor which have been already noticed for
their number and variety. The same vein with its ramifications
cuts through them all but it differs very much in its nature in
different parts. In the limestone and stratified basalts it
often contains lead with quartz and heavy spar [barite]. In the

hard shale it sometimes contains a little copper with calcareous
spar [calcite], and in the coal and soft shale strata it is principally filled with calcareous spar and pyrites.
                              Instance

Figure 8.2. High Force. Davy MSS (15e, 107) 1804. See also
Figure 6.9.

        At the fall of the Tees in Yorkshire, of which I shall show
a representation, there is a considerable vein which passes
through all the strata.
                              Instance
It is here thrown into the picture in water colors. It principally consists of carbonate of lime or calcareous spar, but it
produces a little lead and blende [sphalerite?] in the stratified
basalt. In the shale it contains some quartz, and in the limestone some pyrites.
        It often happens that where the same veins pass through dif-
ferent strata there is a shift or dip, so that the same strata on
one side of the vein are lower than those on the other side, which
would seem to indicate that the original fissure had been produced either by the elevation of one part of the rock, a force
acting from below, or by the depression or sinking down of the
other part.
                              Instance
        Amongst the primitive rocks, the veins most usually found in
granite are those of quartz and schorl, which often contain the
ore of iron, sometimes tin and manganese, but seldom copper.
Veins of calcareous spar or fluorspar seldom occur in this rock,
but when they do occur they almost always contain metallic substances.

The great vein of lead at Strontian in Scotland is worked
in a mountain of granite. The sides of it are principally calcareous spar but with other beautiful substances, amongst which
are zeolites and the peculiar fossil, strontianite, which has derived its name from the spot. Where these crystallized bodies
most abound, there the ore is found in the greatest quantities.
                          Instance
    Veins of the same kind as those found in granite occur in
siliceous schist. But the schist likewise contains often fluorspar, which is usually found in the same vein with tin or copper.
The siliceous schist is generally most productive of tin; the
soft or grey argillaceous schist, of copper. But both these substances are sometimes found in the same vein.
    Quartz rock and the micaceous schist seldom produce metal-
lic veins, at least I believe none have been found in them in
this country.
    Primitive limestone has sometimes been discovered productive
of copper. And in the quartz and calcareous veins that it contains, gold and silver occur.
                          Instance
    Tin has been found in Cornwall in porphyry, but this rock is
seldom very abundant in metallic veins.
    Serpentine is usually intersected by a great variety of
veins, but none of them are metalliferous. The common substances
found in them are steatite and asbestos.
                          Instance
    Amongst the secondary rocks of the third family (the strati-
fied rocks), limestone is above all others productive of the
metals; copper, lead, iron, and manganese are found in it. In
Derbyshire, limestone is the only metalliferous rock, but in the
north of England the stratified basalt, as we have seen, likewise contains lead.
    The crystallized substances which usually form the sides
of the veins in the different secondary rocks are principally
fluorspar, heavy spar, and quartz. Sandstone seldom or never
contains any metals. Metallic veins are never found in loose
shale nor in columnar basalt.
    The columnar basalt is seldom intersected by many veins,
and when they occur they are either calcareous spar or zeolite.
Zeolite indeed often occurs in it in large cavities.
                          Instance
    There are certain general rules with regard to the indi-
cations of metals which will apply equally to the veins in the
primitive and the secondary rocks. Whenever a vein of quartz,
spar, or any other white substance near the surface contains in
its cavities much brown powder or dust, there is always a great
probability that a metallic ore will be found at no great distance. This brown powder is iron in a particular state, which
almost always accompanies tin, lead, and copper.
    A green colour in a vein of quartz or spar likewise gen-
erally indicates that metallic ore is at no great distance. A
dull green substance of this kind in the Cornish and Saxon mines

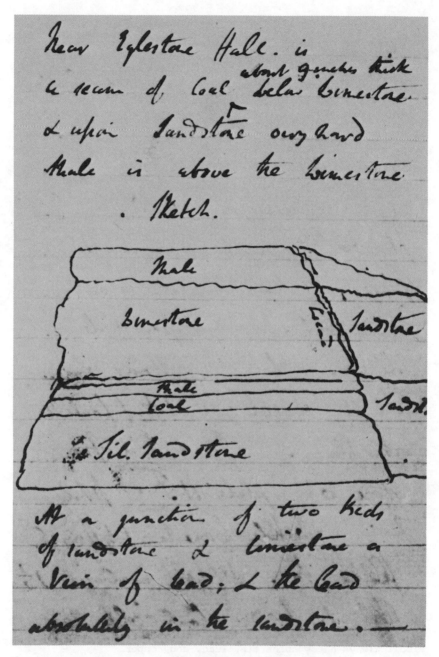

Figure 8.3. Vein near Eglestone Hall. "Near Egleston Hall is a
seam of coal about 9 inches thick below limestone and upon sandstone. Very hard shale is above the limestone. At a junction of
two beds of sandstone and limestone, a vein of lead; and the lead
absolutely in the sandstone." Davy MSS (15e, 145) 1804.

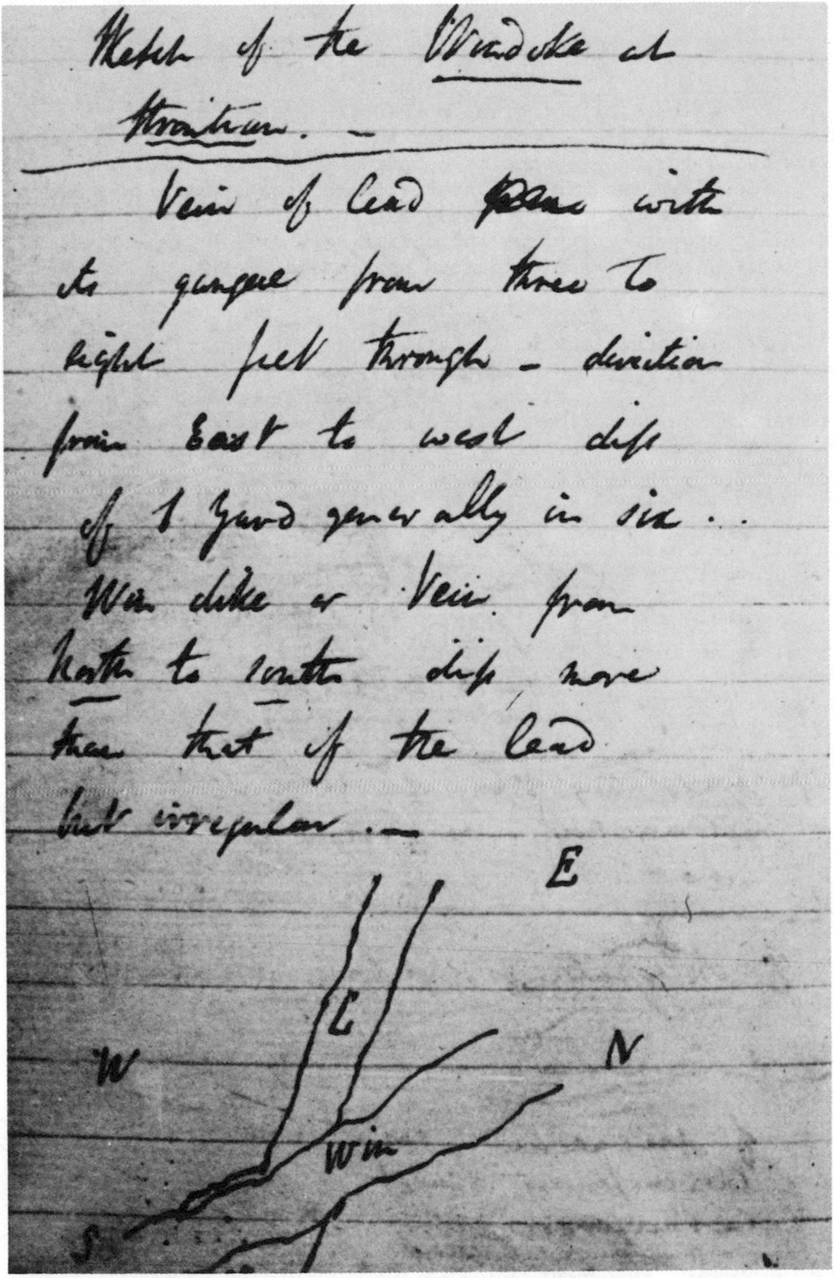

Figure 8.4. Sketch of the Windike at Strontian. "Vein of lead with its gangue from 3 to 8 feet through--deviation from east to west, dip of 1 yard generally in 6. Windike or vein from north to south; dip more than that of the lead, but irregular." (Strontian lies at the extreme left of Figure 5.2.) Davy MSS (15e, 145) 1804.

is often found in the same vein with tin. It is called chlorite.

Instance

But where a bright green powder or crystallized substance is found, it indicates copper and the tint is itself almost always owing to the presence of copper.

There are two kinds of spar which scarcely ever occur in the veins of a secondary country without being connected with metallic ore; they are the rhomboidal spar called pearl spar [dolomite] and the heavy fibrous spar called witherite or carbonate of barytes.

Instance

The various stony and metalliferous veins have different directions in different countries and no universal rule is applicable to them. Many of the great veins that produce gold and silver in South America are said to have a direction from east to west. Some of the principal veins in Hungary and Transylvania run fron northeast to southwest.

But in Britain the greatest metalliferous veins all have their course within a few points from east to west. This is equally the case in the primitive and in the secondary districts. In Cornwall, in Devonshire, and in Argyllshire, in Cumberland, Northumberland, and Durham.

This sketch will exhibit the general directions of the principal veins in Cornwall containing ore. The green represents copper; the blue, lead; the black, tin. From the sketch, it may be seen that the principal veins run very near upon the point of junction between the schist and the granite.

In the secondary country of the north of England, the metalliferous veins, which contain very little of any other metal than lead, have a similar direction. There are however in every part of the island some veins which have a direction from north to south and from northeast to southwest, but they rarely contain any metallic ores and they usually intersect the other veins, which demonstrate that they are of later formation.

These cross veins often contain whin, both in the primitive and secondary countries.

Instance

It often happens that where the cross veins cut the other veins the direction is unaltered, and the vein of the first formation is found on the other side. But it sometimes occurs that where the intersection takes place, there is a considerable change in the positions of the strata, as if the force applied to produce the fissure filled by the cross vein had acted laterally so as to remove one part of the rock containing the first vein to a distance from the other parts.

Instance

This appearance is called by the Cornish miners, "the heaving of the load." And when it occurs in the working of mines, it is sometimes a very difficult task to find the disunited and displaced vein.

It would be little interesting to pursue the more minute details with regard to this subject. No objects can properly

be said to belong to geology which merely present insulated
facts, and the great end of [illegible word] in this science is
to develop some general principles of reasoning or some simple
methods of classing the phenomena so as to guide us by analogy
to new truths or to useful applications.

No branch of practical knowledge can be more simple or more
intelligible than the history of the position and arrangement
of metallic veins. And yet no department of speculative enquiry can be imagined more obscure than the theory of the
causes of their formation and of the laws of their changes.
There have been few writers on minerals, however, who have entered upon enquiry concerning the origin of the metals. Amongst
the alchemists it was a favotite subject. It was examined by
the earlier chemists, and though it has been generally neglected
as a hopeless investigation by the later experimentalists, it has
still been preserved as a favorite subject for discussion and
controversy in the different geological hypotheses.

Agricola, whose name I have already often mentioned, was the
first observer who attempted to give a full explanation of the
production of veins. He supposes that the fissures in which they
are found were partly formed at the time of the production of the
rocks themselves, and partly produced afterwards by the agency of
water. And he conceives that they were filled with stony and
metallic matters by means of the infiltration of fluids saturated
with these different substances.

This opinion, when we consider the time in which it was
stated, may be regarded as singularly clear and very ingenious.
Agricola had seen in many instances metallic precipitations produced from the mineral impregnations in the waters of different
mines, and to him it was yet a problem how far the solvent powers
of water might extend. He had reasoned only from the facts in
his possession, and so far he may be regarded as having reasoned
like a true philosopher.

The celebrated Becher, born at a time when the wonderful
changes produced in bodies by chemical agencies were first beginning to be displayed, concluded that the principal operations of
nature might be imitated by artificial means. He had produced a
small quantity of iron from substances principally volatile. And
he concluded that the metallic veins were formed by vapours which
arose from the interior of the earth and which by acting upon the
stony bodies found in veins transmuted them into different ores.

Stahl, the disciple of Becher, has discussed his opinion in
a work published in 1703, and in collecting facts with regard to
it soon found that it could not be reconciled to the appearances.[4]
This philosopher, as if fatigued with hypotheses and perplexed
with doubt, gave up in his last work all speculation upon the
subject, and rested satisfied with the idea that veins and the
various minerals they contain were created at the same time with
the globe and that they were consequently as ancient as the mountains in which they are found.

Henckel [1678-1744; DSB 6:259][5] of Freiburg, who flourished
at the same time with Stahl, supposes the different veins found

in rocks to have been filled by metallic impregnations formed in consequence of fermentations continually going on in the interior of the different strata. His reasonings were principally founded on some facts which he had observed in the decomposition of pyrites. They were adopted by several of his disciples in the Saxon School of Mines where they were defended with great earnestness for several years.

The idea of Henckel is sufficiently fanciful, but it is perhaps exceeded in singularity by that of Lehmann who attempted to show in his work on the beds of metals, published in 1753 at Berlin, that all the different veins are merely branches arising from one great trunk hidden deep in the bosom of the earth. And he supposes that the substances by which the metals are formed have been elevated in the different branches in the same manner as the sap. is raised in plants, by the effects of heat and of light.

It would be an unprofitable task to examine in detail all the different vague ideas which have been formed upon the subject and which either have not been regarded by accurate thinkers or forgotten almost as soon as they were produced. Amongst the early opinions, I have related only those of celebrated men, and all that they serve to prove perhaps in that when celebrated men amuse themselves with dreaming, they do it almost to as little purpose as common men.

The latest hypotheses that have been formed upon the origin of veins are immediately connected with the two systems of the earth, different parts of which have already been discussed. I shall not take up your time by entering into any minute examination of the comparative merits of the Plutonic and of the Neptunian systems in accounting for these phenomena, for a very transient view will be sufficient to show that both are in their present states inadequate to the explanation of the facts.

Dr. Hutton, reasoning upon the laws of specific gravity, supposes that considerable masses of metallic substances must exist in the interior of the earth. He supposes that the same fire which acting under pressure became the cause of the consolidation of the strata was the agent which brought those metallic substances into a state of fluidity; and he attributes to the same force acting from beneath, the fissures formed in the strata and the projection of the ores into those fissures.

I shall again exhibit the sketch which has been before shown, and it will perhaps elucidate this particular instance better even than the general theory.

The idea of Dr. Hutton is a very bold one, but had it been true, the metals ought to have been found generally diffused through veins, or at least they ought not to have been limited to particular strata. And there is a fatal and a conclusive objection; most of the veins in secondary countries diminish in size as they are worked to a greater depth and, in some instances, though in a calcareous stratum, they have been found to disappear altogether.

Mr. Werner, who upon this occasion may be considered as the representative of the Neptunists (for Mr. Deluc has touched only very transiently upon the subject, and Mr. Kirwan has only lightly examined it in his *Geological Essays*), Mr. Werner, I say, in attempting to account for the origin of veins has adopted an opinion very similar to that of Agricola, though extended and adorned with a variety of new instances. The Saxon professor, in his work lately published upon veins,[6] after discussing the various opinions, endeavours to show that all veins are fissures formed after the rock in which they are found and filled from above. And he asserts that the different matters they contain were dissolved in the same fluid which deposited the primary and secondary strata, and that these substances were precipitated in different crystallized compounds on the sides of the vein.

In a geological point of view, perhaps, there are fewer objections to this hypothesis than to the other, but when chemical affinities are considered, insuperable difficulties arise. We are acquainted with no fluid capable of dissolving the different metallic ores and the substances found in veins; the idea of such a mixed solution is incompatible with the discovered laws of attraction. And even supposing such a menstruum, no cause is assigned why it should lose its active powers and why after being saturated with all the materials of the surface of the globe it should become the comparatively inactive substance water.

In stating that I am not satisfied with Mr. Werner's theory of the filling of veins, I hope it will not be conceived that I am passing a general censure upon his work. I may safely say that the practical observations in it are admirable and worthy of the man who has devoted twenty years of his life to the study of mineralogy. Precision and candour mark all the statements about geological appearances, and where the author has failed in speculation it must not be attributed to a defect of talent but to a want of facts. He has failed in common with many other men of genius who have endeavoured to interpret by the mere imagination what is concealed in nature from the senses.

There certainly appears to be no impossibility in the discovery of a rational theory on the subject, but novel experiments must be invented and new knowledge acquired before the attempt at generalization should be made. Chemistry must first be brought more effectually to the aid of mineralogy, analytical investigations must be multiplied, and those substances which have been principally operated upon in their permanent forms should be examined during the time of their change.

In these objects, there is sufficient to employ the activity of experimental genius for years, perhaps for ages. There is sufficient constantly to keep alive the strongest hopes of discovery, and there is sufficient importance of end to excite the highest ambition of a scientific man.

Lecture Nine

By far the greatest changes produced upon the solid parts of the surface of the globe in late times have resulted from the agencies of volcanic fires. The causes and the effects of volcanoes are subjects which cannot fail to excite curiosity in all minds attached to the knowledge of nature. Their importance in geological theory is great and distinct, for they afford the only known facts of the agency of heat upon rocks which occur in a great scale, and their relations to general theory are numerous and interesting.

Volcanic fires have operated to a great extent in all quarters of the globe, they appear to have occurred in the earliest ages of which we have any authentic records. Pindar, who lived nearly 500 years before the Christian era, mentions the eruptions of Etna, and blending descriptions of this great natural event with poetic fable, he attributes the effect to Typhoeus who, after being defeated by Jupiter in the wars of the gods at [Phlegraean Fields],[1] was supposed in the ancient mythology to have been buried under Sicily. The passage is one of the finest in the odes. I shall read a few lines of it in West's translation.[2]

> Now under sulphurous Cuma's sea-bound coast,
> And vast Sicilia's lies his shaggy breast;
> By snowy Aetna, nurse of endless frost,
> The pillar'd prop of heaven, forever press'd:
> Forth from whose nitrous caverns issuing rise
> Pure liquid fountains of tempestuous fire,
> And veil in ruddy mists the noon-day skies,
> While wrapt in smoke the eddying flames aspire. . .

Figure 9.1. Eruption of Vesuvius from Naples, 18 June 1794. From William Hamilton, "Account of the Late Eruption of Mount Vesuvius," *Philosophical Transactions of the Royal Society of London* (1795), vol. 85, tab. 7, p. 116.

Amongst the Greek historians Thucydides mentions a great eruption of Etna which took place 476 years before Christ, but he gives no minute details.

Of the Greek naturalists Aristotle in his Περι θαυμασιων ματων, or *Concerning Wonderful Reports,* states that he had heard of fires which burnt by night and issued out of the earth in the Aeolian Isles, and he was perfectly well acquainted with the volcanoes of Sicily.

The first eruption of Vesuvius certainly known to have taken place was in the seventy-ninth year after the Christian era, that eruption in which the elder Pliny lost his life. But Diodorus Siculus and Strabo, who were both born before the death of Julius Caesar, were of the opinion that the mountain had burnt in remote ages. That this must have been the case is evident from the circumstance of Pompeian and Herculanean cities, which were destroyed in the first recorded eruption, being principally of lava.

The ancient philosophers seem to have paid very little attention to the natural history of volcanoes. And even amongst the moderns, till within the last forty years, it has been but little an object of study. Sir William Hamilton [1730-1803; DSB 6:83] was one of the first persons who endeavoured to direct the observation of men of science to the subject.[3] His remarks upon Vesuvius, published in the *Philosophical Transactions,* are well known, and his work on the Phlegraean fields equally demonstrates the extent of his knowledge and the refinement of his taste.

Since the period in which the first remarks of our celebrated countryman were made known to this world, the path that he has marked out has been pursued by many philosophical observations. Since that time Spallanzani [1729-99; DSB 12:553] has given many valuable remarks on volcanoes in his work entitled *Travels into the Two Sicilies.*[4] Breislak [1750-1826; DSB 2:439] has added considerably to our geological knowledge of them by his observations on the Campania.[5]

Ferber and Dolomieu have both treated with great acuteness the mineralogy of this branch of science.[6] And if the most instructive of the works upon the subject were to be selected, it would perhaps be the treatise of Dolomieu, *Sur les Isles Ponces,* on the Lipari Islands.[7] In this publication the most minute observations may be found combined with the most enlarged views, and the author has enlightened his philosophical researches by the assistance of the most refined methods of chemistry.

The forms of volcanic mountains are usually conical, and those rocks which constitute their bases and which have not been altered by fire differ very much in their nature in different districts. The strata that appear at the foot of Vesuvius are principally shell limestone. The foundations of Etna are said to be granite and porphyry, and these substances, according to the reports of Humboldt [1769-1859; DSB 6:549], constitute the bases of the great volcanic chain of the Andes, some of the eminences of which are elevated to the enormous height of 4 miles above the level of the sea.[8]

Figure 9.2. Vesuvius after the Eruption of June 1794. The
dotted lines A and B represent the appearance of the summit of
Vesuvius before the eruption. William Hamilton, "Account of the
Late Eruption of Mount Vesuvius," *Philosophical Transactions of
the Royal Society of London* (1795), vol. 85, tab. 8, p. 116.

The lower region of volcanic mountains, even when the sub-
stratum of the soil is lava, are usually exceedingly fertile, and
when a long, continued quiet has prevailed in the interior, vegetation is vigorous upon all parts of the surface that are not
raised above the line of perpetual snow.

In the account given of the crater of Vesuvius by Braccini
before the eruption of 1631, at a time when it had been quiet
for a great number of years, the interior of the mountain is described as in the form of the inverted hollow of a cone, covered
with luxuriant vegetation and abounding in majestic trees, and
in those places where of late years the fiery flood of lava has
boiled up, he describes streams of pure water as flowing down
and cooling the air and affording nourishment to the various
tribes of plants that adorned their banks.

The height of Etna is about 10,030 feet and its circumference no more than 40 miles, and yet so fertile are the soils in
the lower part of the mountain that the population is immense
and the number of the inhabitants of the district is said to
exceed 300,000.

A friend who travelled into Sicily in the year 1803 has
favoured me with a perusal of his notes made upon Etna.[10] They
were written with no other view than that of recalling to his
memory scenes that had strongly impressed him at the moment. I
know that his descriptions are just and they are so vivid that
I am sure I need make no apology for introducing them in illustration of the subject.

In speaking of the lower part of Etna he says the ascent
at first is not very difficult.

> Fields of exuberant fertility smile on every side.
> Sarsaparilla, cinnamon, sassafras, and the coffee plant,

Figure 9.3.   Mount Vesuvius and Mount Somma. This engraving,
which appears in John Davy's *Memoirs of the Life of Sir Humphry
Davy* [2 vols. (London, 1836), 1:503], is based on a crude sketch
made by Davy in March 1815. After ascending Vesuvius and going
around the base of Somma, Davy wrote, "I think there can be no
doubt that the eruption which raised the cone of Vesuvius split
Somma in sunder [sic] and threw part of it off towards the sea.
Thus Vesuvius rises out of Somma." Davy MSS (14i, 142). The re-
lationship between Vesuvius and Somma is very clearly represented
in Figure 9.1.

and saffron all grow here in abundance. It is certain
that no part of the island can vie with the borders of
Etna in verdure, in herbage, and in the luxuriance of
the grain.

The ground is agreeably diversified by the number of
small hills which at different periods have formed an
outlet for the subterranean fires, but which in the long
process of time have become covered with vegetable mould
and by their present abundance atone for their ancient
sterility. Trees so scarce in other parts of the island
spring up here spontaneously and attain a strength and
vigour scarcely to be seen elsewhere. Amongst them peep
out a number of small, country houses and villages, the
inhabitants of which are well formed and healthy. Almost
sinking under the heat of Catania, the freshness of the

air which we here experienced seemed a blessing sent from
heaven, and every object that we beheld appeared warmed
with bounty and filled with life.

The summits of volcanic mountains, when elevated to
any considerable height or after having felt the effects
of an eruption, present a striking contrast when examined
in the same view with the lower regions. Black, and
rugged, and barren, if not covered with snow, they bear
the evidence of having been changed by the force of fire.
Formed of lava and strewed over with pumice stone or with
volcanic dust and sand, they present pictures of which
all the features are wild, and sublime, and more impressive as they necessarily call up in the mind the
imagination of the great natural events by which they
were formed.

Most of the craters or interior cones of the great
volcanic mountains, in their common states, emit smoke
or some kind of vapour, and the lava of which they are
composed is very irregular in its forms and various in
its constitution.

Breislak thus describes the crater of Vesuvius as it appeared in 1800:

> The cone, he says, is cut in an inclined plane having
> its direction from northeast to southwest. The cir-
> cumference of the crater is about 3,000 feet. At the
> bottom is a considerable plain from which vapours of
> different degrees of density almost constantly issue,
> and the sides of it offer the remains of the last lava
> that issued from it in the eruption of 1794.[11]

Instance

Dr. von Troil mentions very similar facts with regard to
the different craters of Hecla in Iceland which, having their
exteriors covered with snow, constantly emit from their interiors
vapours insupportably hot and exhibit walls of lava having the
appearance of black glass.

My friend who I have before quoted spent some hours upon
the summit of Etna in the summer of 1803. I have transcribed
his description of this journey to the edge of the great crater.
Large tracts of snow, he says, led to the foot of it.

> The ascent is prodigiously steep and rendered the more
> difficult by the immense blocks of lava on which you are
> constantly obliged to tread and which never afford a se-
> cure footing. We soon reached the spot whence issued the
> late eruption only about two months ago. The aperture is
> very small; sulphurous steams issue from it in great quan-
> tities and large masses of sulphur appear in its neigh-
> borhood. The ashes, sand, scoria, and lava of which the
> external part of the crater is composed have no consis-

tence. Hence you slip at every step, but in slipping you
make a footing. We had occasion to repose several times,
but by degrees we attained the summit.

The present crater has been formed within the last
century, chiefly by eight great eruptions. Its height,
according to Ferrara [1767-1850], is 1800 feet; the circumference of its base 1200 feet.[12]

The view which presented itself from the top of the
crater was perfectly new, altogether different from what
I had expected. As you have no object with which to compare the hollow, you can form no accurate judgement by
the eye of any of its dimensions. The ground of the bor-
der is so hot as to burn the feet. Vapours issue from
numerous orifices in it. And from the huge abyss itself
arise volumes of thick smoke which darken the face of
heaven and the colour of which is constantly varying.
The sun seen through this smoke sometimes appears perfectly white and sometimes of a blood red colour. We
looked with long and anxious expectancy for a favoured
moment in which the currents of vapour might disappear or
become less dense so as to exhibit to us the interior of
this wonderful laboratory of nature, but we looked in
vain. We rolled down large stones with the hopes of
being able to estimate the depth from the time of their
descent. The ground shook under us as often as they
struck against the side of the precipice, and the sounds
became fainter and more indistinct till they were gradually lost, but we had no reason to suppose that we
heard the noise of their last fall.

Instance

Spallanzani, as he himself relates, was more fortunate as to
the circumstances under which he beheld the crater in September,
1788, and unless he was much deceived he saw the very lowest
part of the aperture.[13]

I sat down, says this author, near the edge of the
crater and remained there two hours. I viewed with as-
tonishment the configuration of the borders, the internal sides, the form of the immense cavern, its bottom an
aperture which appeared like the melted matter which
boiled within, and the smoke that ascended from it. The
whole of this stupendous scene was distinctly displayed
before me, and I shall now proceed to give some description of it though it will only be possible to present the
reader with a very feeble image of a most grand and astonishing sight.

The upper edges of the crater are about a mile and a
half in circumference. The sides are composed of rugged
pieces of lava, form a species of funnel, and abound with
concretions of sal ammoniac. The bottom is a horizontal
plane about two-thirds of a mile round. In this plane

a circular aperture was visible from which issued a large
column of smoke, and when this smoke was impelled to the
side opposite to that on which I was, I can affirm that I
perceived very distinctly a liquid ignited matter which
continually undulated, boiled, and rose, and fell without
spreading over the bottom. This certainly was the melted
lava which had risen to that aperture from the bottom of
the Etnean gulf.

Instance

The general aspect of a volcanic district, even in its most
quiet state, must be highly impressive, but when the subterranean
fires are displayed in their full energy, when they burst forth
from the interior of the earth desolating and destroying, then
the effect must be beyond all comparison the most awful and the
most sublime of the phenomena belonging to our globe.

Multiplied descriptions have been given of the eruptions of
volcanoes. Philosophers have detailed them, poets have painted
them, but language must necessarily fail when applied to such a
purpose. And not even the most perfect delineation of the most
perfect artist could do justice to a combination of circumstances in which feeling, and hearing, and sight are almost
equally concerned; in which the earth trembles, in which the
continued sound of thunder dwells upon the ear, and in which
the eye is constantly dazzled by lightning flashing above and
by liquid fire streaming below.

From the most accurate accounts it appears that before any
great eruption takes place, the mountain for a considerable time
is more than usually tranquil, and one of the most common indications of the approach of the event is a great stillness in the
air, slight tremblings of the earth, and a drying up of the
streams in the vicinity.

A violent earthquake often accompanies the first great
appearance of smoke and flame, and the mountain is often in
combustion for many days before any lava appears. At other
times, however, in cases when great explosions have taken place,
the lava is thrown out at the same time with the smoke and flame.
This happened in the eruption of Vesuvius in 1794, one of the
greatest recorded.

The general progress and extensive effects of this event
were minutely observed and have been accurately detailed by Sir
William Hamilton, and I trust that an abstract of a few of the
most remarkable particulars, taken from his paper published in
the *Philosophical Transactions* for 1794, will not be found tedious.[14] They immediately belong to the subject of the lecture
and will offer some curious elucidations of theory.

The mountain, says the illustrious observer, had been
remarkably quiet for seven months, nor did the usual
smoke issue from the crater, but at times it emitted
clouds that floated in the air in the shape of little
trees. It was remarked that for some days preceding the

eruption a thick vapour was seen to surround the mountain about a quarter of a mile beneath its crater and the sun and moon had an unusual reddish cast. The water at the great fountain of Torre del Greco and of the other springs in the vicinity began to decrease some days before the eruption. On the twelfth of June there was a violent fall of rain and soon after the inhabitants of Resina, situated directly over the ancient town of Herculaneum, were sensible of a rumbling subterraneous noise, and about eleven o'clock at night a violent shock of an earthquake was felt at Naples.

On the fifteenth of June soon after ten o'clock at night, another shock occurred, but not so violent as that of the twelfth. And at the same moment a fountain of bright fire attended with a very black smoke and a loud report was seen to issue and rise to a great height from about the middle of the cone of Vesuvius. Soon after another of the same kind broke out at a little distance lower down and both the openings immediately poured forth lava. Fresh fountains of liquid fire succeeded each other hastily, and in the space of a mile and half, fifteen of them might be numbered and others most probably were hidden by the smoke.

It is impossible that any description can give an idea of this fiery scene or of the horrid noises that attended this great operation of nature. It was a mixture of the loudest thunder with incessant reports like those from a numerous heavy artillery, and accompanied by a continual hollow murmur like that of a roaring of the ocean during a violent storm.

There was an intermission of the eruption at about two in the morning of the sixteenth, but at four, the crater again showed signs of a new action. Lava issued from a new opening on the other side of the mountain facing Ottajano, and in a short time had passed over three miles of ground and overwhelmed and burnt a considerable wood.

The upper part of Vesuvius was again soon involved in clouds and darkness and so it remained for several days. But above these clouds fresh columns of smoke were continually rising till the whole mass had the form of a fine tree stretching to an enormous height in the atmosphere. In that gigantic mass of heavy clouds, flashes of forked lightning were constantly observed so bright as to be visible in the day time.

At about six o'clock in the morning of the sixteenth, the lava had reached the sea after having passed through and destroyed the greatest part of Torre del Greco. It met the waves with a hissing noise, and with loud explosions, and continued some time to flow beneath them.

Instance

Some of the eruptions of Etna have been upon a scale still greater than even this of Vesuvius. In the eruption of 1669 the effects were the most tremendous that have been noticed in any age or in any country. These effects have been minutely described in a letter from Lord Winchelsea to Charles the Second. His Lordship was an eye witness of the event and has related it with much strength of coloring and liveliness of description which even the quaintness and ambiguity of his style cannot obscure. I shall make no apology for extracting a few sentences from the work.[15]

> I accepted, says the author, the invitation of the Bishop of Catania to stay a day with him so that I might be the better able to inform your Majesty of the extraordinary fire which comes from Mount Etna, 15 miles distant from that city, which for the horridness of its aspect, for the vast quantity thereof (for it is 15 miles in length and 7 in breadth), and for its monstrous devastation and quick progress, may be termed an inundation of fire, a flood of fire, cinders, and glowing stones burning with such rage as to advance into the sea 600 yards and that to a mile in breadth, which I saw. And that which did augment my admiration was to see in the sea this matter like rugged rocks burning in 4 fathoms of water and 2 fathoms higher than the waves, with some parts liquid throwing off the stones about them and causing a great and horrible noise in the water, and other parts becoming solid and making a firm foundation.
> When it was night I went upon two towers and I could plainly see at 10 miles distance the fire descending from the mountain in a direct line and the flame ascending as high as the greatest steeple in your Majesty's kingdom and throwing great stones into the air. And I could discern the rivers of fire flowing down from the mountain of a bright red colour, bearing upon them great stones of a paler red. I assure your Majesty no pen can express how terrible it was, nor could all the art or industry in the world have quenched it or diverted its course. In forty days time it had destroyed the habitations of 27,000 persons, made two hills of a thousand paces high each. Of 20,000 persons that did inhabit Catania, 3,000 only did remain. The fire in its progress towards Catania met with a lake of 2 miles in compass, and it was not satisfied by filling it up, though it was 4 fathoms deep, but it had made a mountain of it.

All the accounts that have been published of eruptions in other parts of the world agree in their general features, and from the similarity of the effects of volcanic fires, a strong conclusion may be formed concerning the similarity of their causes. On this subject however a number of difficulties oppose themselves to our reasonings.

The agents concerned in those great subterraneous changes are hidden deep in the bosom of the earth, and it is impossible for us to contemplate them in their active state. Even their effects can only be considered at a distance, and from the grand and awful nature of those effects, it is scarcely possible even for the most philosophical and courageous mind to examine them with coolness and precision. Here analogy must be our principal guide, analogy investigated with caution and assisted by accurate experiment.

Many enquirers into nature, in attempting to account for volcanoes, have supposed that a permanent central fire occupying the interior of the earth is the universal cause of igneous eruptions. This notion is one which it requires little depth of research to adopt and but little acuteness of understanding fully to refute. Heat in all its states is capable of being communicated from body to body, and if a permanent fire had been acting for ages upon the interior of the crust of the globe its effects must long ago have been perceived upon the whole of that surface, which would have exhibited not a few widely scattered volcanoes but one ignited and glowing mass.

An opinion infinitely more probable is that volcanic fires are the result of chemical changes of the intimate agencies of bodies of inflammable matters, producing violent fermentations within the solid strata of the earth. Lemery [1645-1715; DSB 8:172] in 1670 was the first person who brought forward this theory under any precise form, and he founded it upon an experiment of his own invention.[16] He buried two parts of wetted sulphur and one of iron beneath the ground and in a few hours they fermented, heated, and flame arose from the surface with slight explosions. Sulphur, Lemery knew, was one of the common products of all volcanoes and he knew that iron abounded in the materials thrown out by them. His hypothesis was immediately pursued, and these two agents appeared to him adequate for the explanation of the effects. He had not himself been a traveller into volcanic countries, and in his time all the observations made were superficial, transient, and inadequate. It is not the defect of his genius, therefore, if his theory be insufficient to explain the phenomena; it is the defect of the knowledge of his age.

The late researches in experimental science have shown that iron cannot be combined with sulphur in the state in which it is commonly found in metallic veins, or in mines, or in volcanic countries, and therefore no distinct conclusion can be formed from the artificial process on the paste of iron and sulphur with regard to the real cause of subterraneous fires.

Since the progress of the great discoveries in pneumatic chemistry, many guesses have been formed having for their foundation the original principle of the dependence of volcanic effects upon chemical action, but few of them have been produced by enlightened experimentalists. The most sagacious of the chemists of Europe hitherto have attended more to art than to nature. Filled with admiration of the great changes produced

in their experiments, they have principally directed their researches to minute processes for changing the forms of matter
and for applying the various substances belonging to the earth
to the uses and comforts of man, and occupied with those truly
interesting objects they have as yet attended but little to the
natural operations of the globe and to the changes that took
place above and below the surface.

However, these departments of the science offer, whenever
they shall be cultivated with ardour, rich harvests of discovery.
And even in the present infant state of the enquiry the few truths
that have been ascertained with regard to them have principally
resulted from chemical investigations. The chief materials active in volcanic eruptions must be newly combined during the time
of their operation, and such of them as are thrown out from the
mountain, be altered by the agency of the air.

We can, however, reason only upon those substances that are
actually discovered, and though the facts known with regard to
them are scanty, yet they will afford us very good foundations
for theoretical discussion. The only known substance found in
the bosom of our mountains which is capable of spontaneous inflammation from the effects of common, natural agents is pyrite,
and this body is one of the most generally diffused in nature.
It is found in almost all rocks; it forms veins and often occurs
in considerable masses. Now the mere agency of water is sufficient to produce such a fermentation in loose masses of pyrite
as to occasion their ignition.

Of this truth there are numerous instances. I will take one
recorded by Dr. Plot in the history of Staffordshire.[17] A man
called Wilson, says the author, who lived at Ealand in Yorkshire,
piled up many cartloads of pyrite in a barn of his own for some
secret purpose, perhaps to extract the gold. The roof being
faulty and admitting rainwater to fall copiously in amongst
them, they first began to smoke and at last to take fire and burn
like red hot coals so that the town was considerably disturbed
and alarmed.

During the fermentation of pyrite with water, sulphur and
the same species of inflammable air which is given out in the
steams of volcanoes are evolved. But there are other products of
eruption which cannot be generated merely from such a cause.
The only operation by fire which can take place on a large scale
and in which sal ammoniac, and bitumen, and fixed air or carbonic
acid (all of which are common volcanic products) can be evolved
is in the distillation or combusion of pit coal.

To those persons who have not been accustomed to consider
the general series of chemical phenomena, this idea may seem to
need experimental proof. It would be easy to offer many but I
shall content myself with one. Included in the top of this model,
there is a mixture of pyrite and coal with a little oxymuriate
of potash, a substance that will speedily supply to the pyrite
oxygen, the principle necessary to its inflammation which water
would only supply slowly. When I touch the mixture with a
little sulphuric acid it will heat, crackle, and emit smoke.

This must not be considered as an imitation of volcanic fire, for then it would be ridiculous, but as an illustration of the considerable active powers of matter. And I shall have occasion to offer more illustrations of the same kind in the next lecture.

If we can conceive that the principal agents in the excitation and preservation of volcanic fires are pyrite and pit coal, acted on by water and air, all the appearances would admit of an easy explanation.

Sir William Hamilton has often remarked that in common seasons, Vesuvius spews out most smoke and ashes when the tides are high and the wind blowing towards the shore, which seems to imply some subterraneous communication between the interior of the volcanic crater and the sea. The drying up of rivers and springs likewise before great eruptions seems to indicate that the water has seome share in the effect.

Now assuming that immense masses of pyrite existed in volcanic mountains, diffused through beds of coal and in contact with limestone, besalt, and other rocks, and supposing that they were acted on by water flowing in upon them from some new orifice accidentally made, the first effect would be a production of heat. And this heat not being carried off by any current, air would constantly accumulate and increase the general series of changes. The elastic fluid evolved, becoming of a high temperature, would in consequence of its expansive power make strong efforts to escape by acting upon the incumbent strata and would produce trembling and agitations in the surface.

In cases in which a crater already existed, the elastic fluid would be discharged through it in great explosions and would carry with it aqueous vapour and bituminous and sulphurous smoke. As the elastic fluid would be unequally discharged, common air would be occasionally pressed in to supply its place, and a current would be established capable of inflaming the combustible matters even in the most profound parts of the mountain. With this effect the phenomena of ignition would begin.

The sulphur, coal, and bitumen in combustion would tend to fuse the compound stones in contact with them. Carbonic acid gas would be produced, and this and the other elastic vapours would necessarily tend to elevate portions of the melted mass into the upper part of the crater. The column of ascending smoke would in consequence be inflamed, the lava would be poured forth, and ignited stones, projected into the air.

With every chemical change, there is likewise a change in the states of electricity of bodies. And where such a series of alterations was produced, the electrified vapours would necessarily give off their fluid to the surrounding conducting surfaces, and the awful phenomena of the ignited exploded matter would be rendered more sublime by lightnings and thunders.

I venture to give this theory as an imperfect sketch. At present, there is no time for a full discussion of it. On next Thursday I shall resume the subject, and till that period I shall likewise defer the consideration of the general effects of earthquakes and volcanoes in the order of nature.

I already anticipate a number of objections to the theory, but before it can be fairly judged of, the minuter facts belonging to it ought to be considered. At first view it will perhaps appear too complex. Accustomed constantly to look for simplicity in nature, the mind can scarcely bring itself to believe that one great phenomenon can be produced by a combination of minute causes. Yet this, as will be shown, often happens in the common and general series of natural events. Powers that have been operating silently and unperceived for ages sometimes produce their ultimate effects in moments. The most diversified of the properties of bodies are often called into activity for one end. And though the laws that govern the material world are few and simple, the agencies that they direct are many and complicated, and as various as the external appearances they produce.

Lecture Ten

The theory of volcanic fires stated in the last lecture certainly needs support and defense, but it ought to be remembered that on such a subject no more can be hoped for than a probable explanation of facts. On phenomena of so impressive a character it is scarcely possible to avoid indulging in speculation. Some hypothesis must be formed, and the greatest merit that a hypothesis can possess is that of accounting for general appearances.

That the principal products of volcanoes are such as would be generated by the spontaneous accension [i.e., kindling] of pyrite from the action of water and by the combustion of coal affords a strong argument in favour of the dependence of eruptions on these causes. And though an objection naturally arises that powerful subterraneous fires often occur in mountain chains in which no coal has been found, yet this objection is founded rather upon our ignorance than our knowledge. For the volcanic nature of the mountain implies something peculiar in its internal constitution, and as carbonaceous matter has been found in small quantities in the granitic rocks of Saxony, there is no reason why it should not likewise exist in the porphyritic rocks of Sicily.

At least pyrite, which is assumed as the prime agent, abounds in primitive countries. It is found for instance in immense quantities in the mines at the feet of the American Andes. Pyrite has often been discovered imbedded in substances thrown out by Etna, and I have seen it in primitive rock which must have been evolved from a very deep part of the crater of Vesuvius, and it is abundant in the Solfatara.

Allowing that pyrite is one of the substances active in volcanic fires, it may still be asked why these fires are not much
more frequent, as the substance is stated to be one exceedingly
abundant in nature. But to this question the reply is obvious;
a combination of circumstances is necessary to produce the effect.
The presence of water is essential, and not merely this, but the
pyrite must [also be] of so loose a texture as to be permeable to
the fluid and must be in very considerable quantities, so that
the heat evolved may be accumulated and constantly increased.

There are extensive veins of pyrite in the different primitive rocks of Cornwall, but they are fortunately of a compact
species and imbedded in hard stone which is impenetrable to
water; hence they can undergo no change.

But in veins of pyrite which are in a loose state of aggregation and which are imbedded in soft strata that are permeable
to water, chemical alterations very often occur. Thus we are
told by Mr. Stevens[1] in volume 52 of the *Philosophical Transactions* that in the month of August 1751, the pyrite in the cliffs
near Charmenter in Dorsetshire being wetted by rain and the
spray of the waves, took fire, and first began to smoke, and
then emitted a strong flame. The effect was imitated by the
author who collected about 100 pounds of the wetted pyrite and
sprinkled it occasionally with water, the consequence of which
was that in about ten days time they grew hot, soon after caught
fire, burned for several hours, and then fell into dust.

A similar circumstance is mentioned by Mr. Williams[2] in his
work on the mineral kingdom, and this is the more worthy of being
noticed as the accension of the pyrite was actually the cause
of the inflammation of a coal stratum. He states that, in Dysart
Moor, the coal which abounds in pyrite has undergone a spontaneous
combustion. Sometimes flames are seen in the night and a black
smoke in the day; and the inhabitants say that, at the approach
of storms, dreadful hissing noises are heard from the holes and
caverns, with a considerable discharge of flame.

Many other accounts of analogous facts might be brought forward and the phenomenon is not very uncommon in the coal districts.

These are specimens of a burnt clay stratum from Burkwood
Common near Altofts which owe their present state of induration
to the combustion of a coal stratum.

Instance

The circumstances of the combustion are not known as no accounts
are upon record, but its effects must have been very extensive.
And when the texture of these substances is considered, and the
strong analogy of some of them to lava, it will be easy to conceive that the same causes operating upon a larger scale and producing their effects within the form of the earth would be fully
adequate to the production of a volcanic eruption.

I mentioned in the last lecture the observation of Sir
William Hamilton on the effects of tides and high water on
Vesuvius. And there are many other facts which seem to prove a
general connection between volcanic eruptions and a subterraneous

agency of water. Thus most of the great volcanoes known are
placed at no considerable distance from the sea, and water in a
number of instances has been thrown out by mountains at the time
of the greatest activity of internal commotion. In the eruption
of Vesuvius of 1631, a great number of vineyards were destroyed,
and a number of lives lost by the torrents of water which issued
down the sides of the mountain and which had every appearance
from their direction of having been thrown from the crater.

And eruptions of water and of mud have in several instances
taken place from Etna. Thus in 1795 an enormous quantity of
aqueous vapour issued from the crater with great explosions, and
being condensed on coming in contact with the cold atmosphere,
it streamed down the sides of the mountain in torrents, the force
of which are irresistible.

Whether the action of water upon inflammable bodies be admitted or rejected as the primary cause of the subterraneous
heat of volcanoes, these phenomena prove at least that it is
very often active in the general order of the operations. The
thick clouds, the heavy rains which in all cases follow eruptions,
demonstrate that a considerable portion of water rendered elastic
by heat is evolved, and there seems to be no agent more adequate
than aqueous vapour for producing the tremblings of the ground,
the convulsions of the strata, and the violent earthquakes which
always form a part of the series of volcanic appearances.

The powerful effects of steam even in its common states are
well known and they are exhibited in the greatest of mechanical
inventions, in the steam engine. But when this substance is
highly heated, its elasticity is increased in a ratio rising with
the temperature. At a red heat, there is every reason to believe
that its force would be more than 1,000 times greater than at the
boiling point, so that the igneous vapour disengaged in volcanic
eruptions must in all cases be capable of exerting a very powerful agency upon the superincumbent rocks, and its influence, even
if not preeminent, must considerably aid the effects of the various permanently elastic fluids and other volatile substances
which are disengaged.

An instance of the very great mechanical force of steam when
highly heated occurred about four years ago in Coalbrookdale in
Shropshire. In a great flood that took place, some water accidentally flowed into an iron furnace containing several tons of
melted metal. An explosion instantly occurred so loud as to be
heard at the distance of 20 miles. All the houses surrounding
the furnace were violently shaken as if by an earthquake and the
whole of the iron propelled in its liquid state into the air so
as to form a column of fire of the most brilliant whiteness and
supposed to be 1,000 feet in height.

Many of the explosions in chemical experiments are likewise
produced by highly heated steam. This is particularly the case
in two powerful compounds called fulminating silver and fulminating gold. When these substances are made to detonate by heat
or friction, water and a small quantity of permanently elastic
fluid are formed, but the great effect is owing to the water

which being highly heated in the process exerts an amazing expansive force.

Instance

There is one circumstance in the general hypothesis that has been states which agrees very well with the facts. The presence of the atmosphere is not supposed to be essential in the first stage of volcanic fire, but it is still considered as connected with its ultimate effects and with its continued duration. And we have seen that volcanic eruptions have often burst forth in the sea, a case in which it is scarcely possible to conceive that air can have been present or that the first process of ignition could have been a process of common combustion.

But in permanent volcanoes, there is often a distinct dependence of the effects upon the action of the atmosphere. In Stromboli, for example, which has been constantly burning for more than 2,000 years, there seems to be a communication between the subterraneous fire and the atmosphere by means of caverns which must be principally situated at the south side of the mountain. The flame is much more intense and the eruptions much more vigorous when the wind blows in a southerly direction, and if from any circumstance the currents of air are interrupted, the lava ceases to boil up on the crater, and smoke [ceases to ascend] up the apertures connected with the atmosphere.

This is a view of Stromboli painted from a plate in Spallanzani's *Voyage to the Lipari Islands*.[3]

Description
Reasoning

Hitherto one series of causes only has been assumed as essential for producing the various effects that have been described, but it is equally easy to conceive that many others may interfere. It is not impossible that beds of nitre, of common salt, of manganese, and of various metallic ores may be active in the operation. Muriatic acid and sulphuric acid have both been found in the fumes from volcanoes, and manganese exists in several lavas. And as is well known, in all artificial experiments of the action of muriatic acid upon manganese, or of sulphuric acid upon salt and manganese, an elastic fluid is produced capable of burning inflammable substances without the assistance of foreign heat, and even of dissolving the metals with all the phenomena of ignition.

Instance

And similar effects would occur if this gas were disengaged at the bottom of a fluid in contact with combustible matter.

Instance

Phosphorus, Oil, or Bitumen, or Oil of Sassafras.

I have already mentioned the electrical appearances occurring in the atmosphere during volcanic eruptions, and there is every reason to believe that the disengagement of electricity, or rather the destruction of its equilibrium, must likewise take place to a great extent below the surface, where it would occasion the most rapid changes. Second, we know from various experi-

Figure 10.1.   Views of Stromboli.   From Lazzaro Spallanzani,
Voyages dans les deux Siciles, 6 vols. (Bern, 1795-97), vol. 2,
frontispiece.

ments that, by the mutual chemical agencies of solids and fluids
upon each other, an unlimited electrical power is generated, and
even in the mere heating and cooling of bodies there is often a
change in their electrical states.   Thus sulphur after being
melted is found strongly electrified.

The electrical fire evolved by the action of compounds of
sulphur and of acids upon the metals has often been exhibited
in galvanic experiments in this theatre.   The instances may
however be new to a part of the audience.

Instance

In these arrangements no agents are concerned but plates
composed of two metals alternating with the fluid.   And in the
chemical agencies with which metals are concerned in the moist
strata of the earth, it is not unlikely that something analogous
may occur.   At least it is certain that such a power must be constantly developed in nature, though its effects may be blended
with those of other causes and are seldom or never exhibited to
us in their simple forms.

No certain knowledge has as yet been gained of the highest
degree of the heat of subterraneous fires, but from their effects
upon different rocks there is no reason to suppose them of a
higher degree of intensity than the fires artificially excited
in our common furnaces.   A number of lavas contain unaltered
feldspar, or feldspar mixed only at the edges, and this substance

which is capable of being rendered liquid in strong white heat
equal to sixty degrees of Wedgwood. But it necessarily follows
from the variety of circumstances of ignition that the power of
heat must differ in different cases, and that agency of it which
fused the basis of pumice stone must be considerably more intense
than that which forms common lava, for this substance is much
more difficult to fuse by artificial means. Yet in the volcanic
fire, its liquidity must have been such as to render it permeable
to the bubbles of elastic matter which seem to have produced the
cavities in it and to have occasioned its porous texture.

Dolomieu, from finding in several instances granite and porphyry apparently unaltered in lavas, supposed that the matter
producing their liquidity was a bituminous or sulphureous substance which burnt during their exposure to the atmosphere. But
this idea, though the production of a very acute mind, seems to
have met with no support from the other philosophers who have
most studied the phenomena of volcanoes.

The lava is said to be poured out from the crater as an
ignited mass. It flows on like red hot metal in fusion and
yields no flame except when it passes over and destroys vegetable or animal substances. In itself it is said to exhibit
no marks of combustion, and there is every reason to believe that
it is composed of an assemblage of the constituent parts of stones
in chemical union and in solution by fire, and assisted in their
relative affinities by the presence of a small quantity of alkaline matter.

As yet no accurate observations have been made upon the
crystallization, by slow cooling, of lavas of which all the parts
have been fluid. This, however, is a subject of high interest
and would probably afford some important conclusions. I mentioned in a former lecture the very curious results gained by Sir
James Hall and Mr. Gregory Watt in their experiments upon the
fusion and consolidation of basalt.

In the case of lavas, nature must have exhibited upon a
great scale all the variety of effects which heat is capable of
producing in its most diversified modifications. And an accurate history of the productions of volcanoes and of the various
strata of stony matter evolved from them would be one of the
greatest and one of the most instructive works in natural science
that could be accomplished, and one that would above all others
point out the limits and the objects of geological theory.

With regard to the various hypothetical opinions that have
been advanced, I hope it will not be conceived that I have placed
any indiscreet confidence in them. In entering upon speculation,
the principal object in my view was to develop the general facts
known with regard to volcanic eruptions, in a connected and analogical order, and to point out the probable relations of their
effects to certain known causes. The perfect explanation of
these great natural phenomena may probably long remain unknown,
as substances and powers of which we have at present no conception may possibly be called into activity in producing them. And
in that great laboratory of nature concealed beneath the surface

of the earth, agents which as yet have never been developed in
the experiments of the philosopher may exert their powers, and
even the same bodies operating under other circumstances may produce effects wholly unknown in our artificial processes.

But still the spirit of active enquiry might not be de-
pressed by such obstacles. Other objects of speculative research
which in former times were equally obscure and equally difficult
have been enlightened by the discoveries of modern science. And
the investigation of natural causes is always a happy exercise
for the human understanding, not a gratification of idle curiosity, but of the love of useful knowledge. And the development
of truths of this kind is of the highest interest, displaying at
the same time the talents of man, the majesty and variety of
nature, the wisdom and perfection of the laws of nature.

Volcanoes when superficially examined appear rather as
accidents than as orderly events in our system. But when they
are accurately considered, it will be found that their effects
are not unimportant in the economy of things and that they bear
a distinct subservience to the general harmonious series of natural operations. This subject is one to which many experiments
of analogical reasoning might be abstractedly applied, but the
discussion of it is properly connected with the general enquiry
concerning the laws by which the surface of the globe is preserved in a present state, and with the phenomena of the effects
of natural agents in producing the decomposition of rocks and the
formation and renovation of soils.

By the agency of the solar heat and of air, a series of
changes is constantly taking place upon the solid parts of the
surface of the globe, and these changes, though in the common
course of natural events they are slowly and quietly produced,
are nevertheless of the greatest importance in rendering the
earth habitable. Vegetables and animals derive their support in
consequence of a continued action of the various elements belonging to our globe, in consequence of modifications in the forms of
matter, which though almost infinitely diversified yet still uniformly blend to produce a great result.

It is not by sudden changes and by great and impressive
events only that the forms of rocks and of strata are modified
and altered. For by the action of water, or air, and of heat,
a series of slow and of gradual changes is constantly taking
place, which are absolutely connected with the arrangement of
the materials of the surface of the globe, and which are absolutely essential to the preservation of the living beings that
inhabit it. By changes of temperature, by the effects of moisture and of the oxygen of the atmosphere, even the most solid rocks
in the long course of years undergo a certain degree of decomposition and become subservient to the production of finely divided
earthy matter capable of forming a receptacle for vegetable life.

The primitive rocks, as I mentioned in a former lecture,
in general are altered with much more difficulty than the secondary rocks, but when exposed under proper circumstances to the
long continued action of the common natural agents, they all

finally yield to [nature's] powers, though in different degrees
and with very different phenomena.

All our rocks, both primitive and secondary, are more or
less absorbent of water and have in fact a chemical affinity for
it. This is a piece of limestone, apparently dry. Another piece
of precisely the same kind and equally free from loose moisture
is now acted upon in this apparatus by heat. Steam is given off
from it. The case would be similar with any other rock, but
slate would produce more water than limestone, and limestone
more than granite or porphyry.

The moisture included in stones, or adhering to their surfaces when expanded by heat or when converted into ice, becomes a
cause of their decomposition, and by continued changes of volume
in consequence of changes of temperature, the texture of [the
stones'] surfaces is gradually altered and rendered loose and
brought nearer to the nature of soil. There are few rocks that
do not contain a certain proportion of iron not fully combined
with oxygen, or pure air, and this metallic substance by its action upon the atmosphere often becomes a cause of the superficial
alteration both of primitive and secondary rocks. This is serpentine.

Instance

This is basalt.

By the immediate action of rainwater merely, likewise many
rocks undergo decomposition by this agent, calcareous matter
is dissolved, and feldspar of certain granites decomposed.

Instance--Granite

And by these operations which take place very slowly, great effects are continually producing in nature. Hence rivers are constantly transporting small quantities of solid matter; hence
there is a constant diminution of the stones forming their beds.
Sand and mud are produced by their [erosion] and the mountain
torrents carry down a finely divided earth to supply the waste
occasioned by the efflux from the flat countries and valleys.

The stones formed by volcanic fires are above all others
liable to decompose to become an excellent soil from the operation of these natural powers. The oxide of iron contained in
abundance in lavas is an immediately acting cause of their decomposition. The lavas formed from basalt are the most liable
to alteration and after these follow such as are formed from
porphyry and such as contain calcareous earth. But whatever may
by the nature of a volcanic country, it seldom happens that more
than a few years are required for the production of a soil, and
by the agency of heat, of water, and oxygen, the primitive earths
gain that state of aggregation which renders them most proper for
the support of plants.

In whatever way a soil is formed, whether by the decomposition of primitive, secondary, or volcanic rocks, a short period
only elapses in the common course of natural events before it is
made the abode of vegetable life. As soon as the thinnest stratum of earth appears, the seeds of mosses, grasses, and of
heaths, which are wafted by the wind over the surface, find in it

a resting place. They drink in the dew, they convert the air into nourishment; their decomposition affords food for a more perfect species of vegetable, and at length a mold is formed in
which even the trees of the forest can fix their roots and which
is capable of rewarding the labours of the cultivator.

The greater part of the surface, the primitive and secondary
strata belonging to our globe, must have become the beds of vegetation in very early times. But since that period the materials
of their soils have been probably often changed and renovated
even in cases when they have remained in their natural state.

The fertility of soils is determined in a great measure by
climate, by situation, and their relation to rivers and moisture.
The most productive in different parts of the globe and in different latitudes differ very much in composition. But in general
it may be said in considering their chemical nature that the soils
of volcanic countries are by far the most fertile. They contain
usually oxide of iron and calcareous matter; their texture is
equable and of a proper degree of permeability with regard to
water. And carbonaceous matter, a substance capable of becoming
the food of plants, is sometimes found in them.

Calcareous soils are generally the most fertile of the
primitive and secondary soils, and some of our marls produce al-
ternately crops of wheat and beans for years together without
manure. The soils in which mild magnesian earth is found are
likewise productive, and the lands situated upon schist, particularly when their colour is red which indicates the presence of
much oxide of iron, are in general more profitable than the
granite soils.

In the same district and in similar situations a most important difference may be often observed in the nature of the
vegetables produced upon soils differing in chemical composition.
Thus the sandstone hills in Derbyshire produce only heath and
peat moss, whilst the limestone hills are usually covered with
short grass and afford a pasturage for sheep. And a similar difference is observed between the uncultivated granite and schistose hills of Devonshire and Cornwall and the chalk strata of the
south downs of Dorsetshire and Wiltshire. The primary ends of the
decomposition of rocks as far as they can be traced are the formation and preservation of soils. And when these ends have been
attained, the vegetable tribes which cover the surface defend the
strata below from a new action of the elements and prevent to a
great extent any novel decomposition from taking place. By their
living powers they accumulate matter from the waters that come in
contact with them and the atmosphere surrounding them, and they
prevent to a great extent, by the agency of their roots, the
light and finely divided earth from being carried down by rains
or scattered by winds.

The conservation of soils is likewise assisted by many other
causes, and the earthy and ferruginous particles dissolved by
water impregnated with carbonic acid are often deposited again in
consequence of the dissipation of the elastic substance or of the
evaporation of the water which held them in solution. And in

these cases they become a principle of union and of new aggregation to the sands or the earths in which they rest. In this
manner tufas and stalactites are formed.

Instance

And on a new change in the state of the oxidation of iron, the ce-
menting powers of pozzuolana and of other volcanic ashes depend.

Amidst the changes produced, during the long succession of
ages, in the forms of the solid parts of the earth, there appears
to be still the same fundamental relation between them in their
existing states and in their application to the purposes of life.
The small quantities of solid matter carried from the land into
the sea are partly replaced by the accumulation of sand and
finely divided earth upon the shores. The diminution of the size
and height of the great mountains is so slow in the common course
of events as to be scarcely ascertainable by human observation.
And the waste that takes part in particular rocks and strata in
one part of the globe is probably fully compensated for by the
elevation of new hills and mountains by volcanic fires in other
regions.

It has been computed that there are at present at least 120
volcanic tracts in the different quarters of the earth. Most of
the land in the European archipelago seems to have been raised
from the bottom of the sea by igneous eruption, and the remains
of late volcanic agencies are distinct in the Antilles and in
many of the groups of isles in the Indian seas. Thus the earthquake and the subterraneous fire have their uses in our system.
They at first terrify and destroy, but a few years only pass away
and their desolating effects disappear; the scene blooms with the
fairest vegetation and becomes the abode of life.

It is not, however, merely by the agencies of inanimate matter that the great equilibrium is preserved in nature. Organized
beings likewise are continually active (without referring to the
labours of civilized man). Great effects will be found to be
produced by many of the inferior classes of animated creatures.
Insect tribes in the bosom of the ocean, the most insignificant
in their individual powers, by their united agencies are enabled
by a blind but wisely and wonderfully directed instinct, to
accumulate solid matter and to raise it above the waves. And
many of the flat islands under the tropics are merely coral rocks
covered with soil. Such rocks are abundant in the Pacific Ocean
and are said to be constantly increasing. It would be an endless
talk to detail all the operations by which the beautiful cycle
of terrestrial events is preserved in a uniform order. The
arrangements of matter are constantly modified; its essence,
continued inalterable.

Amidst the various infinitely diversified changes of things,
nothing can be said to be accidental or without design. Even the
most terrible of the ministrations of nature in their ultimate
operation are pregnant with blessings and with benefits. Beauty
and harmony are made to result from apparent confusion, and all
the laws of the material world are ultimately made subservient
to the preservation of life and the promotion of happiness.

Reference Matter

Notes

INTRODUCTION

1 The first text to appear was by Robert Bakewell, *An Introduction to Geology* (London, 1813), and it was followed by two
   others before 1820, all of which appeared in several editions:
   William T. Brande, *Outline of Geology* (London, 1817); and
   George Greenough, *A Critical Examination of the First Principles of Geology* (London, 1819). For a discussion of the
   founding of the Geological Society of London, see M. J. S.
   Rudwick, "The Foundation of the Geological Society of London:
   Its Scheme for Cooperative Research and its Struggle for In-
   dependence," *British Journal for the History of Science* 1
   (1963): 325-55.
2 Humphry Davy, *The Collected Works of Sir Humphry Davy, Bart.*,
   ed. John Davy, 9 vols. (London, 1839-40), 8: 153.
3 Ibid., p. 180*n*.
4 William Harris, *A Catalogue of the Library of the Royal Institution of Great Britain* (London, 1809).
5 The Archives of the Royal Institution of Great Britain in
   Facsimile. Minutes of the Managers' Meetings, 1799-1900, 14
   vols. (London, 1971). For the lectures of 1806, see vol. 4,
   p. 129; for those of 1808, see ibid., p. 319. The Managers'
   Minutes do not mention any geology lectures for 1809, but the
   Microcosm of London states that the lectures "for the present
   year, 1809" include a course on geology by Humphry Davy.
   Microcosm of London, or London in Miniature, 3 vols. (London,
   1808-11; reprint ed., London, 1904), 3: 37. For the London

lectures of 1811, see *Philosophical Magazine* 37 (1811): 392-
98; 465-70; for the circumstances of the Dublin lectures,
see John Ayrton Paris, *The Life of Sir Humphry Davy* (London,
1830), pp. 218-19 of the single volume edition.

6 The few lectures included are in volume eight.

7 *Philosophical Magazine* 9 (1801): 281-82.

8 Quoted from a letter of Samuel Purkis describing Davy's first
course of lectures, in Paris, *Life*, p. 90. For a general discussion of the Royal Institution audiences, see George A.
Foote, "Sir Humphry Davy and his Audiences at the Royal Institution," *Isis* 43 (1952): 6-12.

9 See Robert Siegfried, "Davy's 'Intellectual Delight' and his
Lectures at the Royal Institution," a paper delivered at the
Davy bicentenary held at the Royal Institution in December,
1978. Forthcoming.

10 We have reprinted Davy's "Introductory Lecture for the Courses of 1805" from Davy, *Works*, 8: 155-66, making minor
changes in spelling and punctuation where required for consistency.

11 The biographical account given here is based chiefly on the
two primary biographies of Davy, both written by men who knew
him personally: Paris, *Life*; and John Davy, *Memoirs of the
Life of Sir Humphry Davy*, 2 vols. (London, 1836).

12 Paris, *Life*, p. 14.

13 Ibid., pp. 10-11.

14 Royal Institution Davy MSS, Box 1, document 137.

15 A few letters from Gregory Watt to Davy are in the Royal Institution Archives.

16 Thomas Beddoes, "Observations on the Affinity between Ba-
saltes and Granite," *Philosophical Transactions of the Royal
Society of London* 81 (1791): 48-70.

17 Note from Thomas Coulson, a childhood friend of Humphry Davy.
Royal Institution Davy MSS, Box 1, document 84. The work
Coulson refers to is William Enfield, *The History of Philosophy . . . drawn from Brücker's Historia critica philoso-
phiae,* 2 vols. (London, 1791).

18 Copy of a letter from Gregory Watt to Davy, 13 April 1799.
Royal Institution Davy MSS, notebook 9, unpaged.

19 Humphry Davy, *Researches Chemical and Philosophical: Chiefly
concerning Nitrous Oxide, or Dephlogisticated Nitrous Air,
and its Respiration* (Bristol, 1800). See also Davy, *Works*,
vol. 3.

20 In a letter to his mother, dated October 11, 1798 and published in Anne Treneer, *The Mercurial Chemist, A Life of Sir
Humphry Davy* (London, 1963), p. 31.

21 From his 'Sketches of Contemporaries' written near the end of
his life in a personal notebook. It has been published by
J. Z. Fullmer, "Davy's Sketches of his Contemporaries,"
Chymia 12 (1967): 131-32.

22 Quoted in T. E. Thorpe, *Humphry Davy, Poet and Philosopher*
(London, 1901), p. 55.

23 Quoted in J. E. Stock, *Memoirs of the Life of Thomas Beddoes, M.D.* (London, 1811), p. 176.
24 Paris, *Life*, p. 77.
25 Humphry Davy, *Elements of Agricultural Chemistry, in a Course of Lectures for the Board of Agriculture* (London, 1813). The Board of Agriculture was created by Parliament in 1793 and modestly funded by the government until its dissolution in 1822. It possessed no policy-making function; its duties were to encourage improved agriculture, coordinate the work of private agricultural societies, and publish useful information.
26 Martin Rudwick, "Hutton and Werner Compared: George Greenough's Geological Tour of Scotland in 1805," *British Journal for the History of science* 1 (1962): 118.
27 Paris, *Life*, pp. 101–2.
28 Managers' Minutes, vol. 3, p. 304, entry for 18 June 1804.
29 Ibid., vol. 4, p. 8.
30 Ibid., p. 9. The Managers' Minutes for 11 January 1805 indicate that twelve lectures were projected originally, but on 1 April the managers requested that Davy omit his lectures during Passion Week and Easter Week (ibid., p. 53). Because Easter fell on 21 April in 1805, Davy could have completed a ten lecture sequence between 7 February and 11 April without any interuption. Ten lectures was also the number projected for the geology course of 1806 (ibid., p. 129).
   In introducing the publication of the four lectures of this course whose manuscripts are in the possession of the Royal Geological Society of Cornwall, Alexander Ospovat conjectures that the 1805 course contained only seven lectures, and that the other three were added only for the course of 1808. ["Four Hitherto Unpublished Geological Lectures Given by Sir Humphry Davy in 1805," with Introduction and Notes by Alexander M. Ospovat, *Transactions of the Royal Geological Society of Cornwall* 21 (1978): 4]. The most specific evidence against this suggestion is Davy's statement at the end of Lecture Nine that he will resume discussion of the topic of volcanoes "next Thursday." Only the 1805 lectures were given on Thursdays; both those of 1806 and 1808 were delivered on Wednesdays.
31 Quoted in Paris, *Life*, p. 131.
32 Davy, "On Some Chemical Agencies of Electricity," *Works* 5: 54.
33 John Davy, *Memoirs*, 1: 436.
34 Davy, *Works*, 6: 344–58.
35 Humphry Davy, *Salmonia: or Days of Fly Fishing* (London, 1828); see also Davy, *Works*, 9: 1–205; and *Consolations in Travel, or The Last Days of a Philosopher* (London, 1830), see also ibid., pp. 207–388.
36 Introductory Lecture, p. 3.
37 Ibid., p. 4.
38 Ibid., p. 9.
39 Lecture 3, p. 45.

40   Ibid., p. 43.
41   Lecture 4, p. 54.
42   Lecture 2, p. 28.
43   Ibid., p. 22.
44   Lecture 3, p. 39.
45   Ibid., p. 40.
46   Lecture 4, p. 51.
47   Lecture 3, p. 44.
48   Lecture 4, p. 57.
49   Lecture 5, p. 68.
50   Ibid., pp. 69–70.
51   Ibid., p. 70.
52   Ibid., p. 68.
53   Lecture 6, p. 75.
54   Ibid., p. 82.
55   Ibid. Hornblende as used in the early 19th century was a catch-
     all term which included the dark minerals of basalt now recog-
     nized as mostly pyroxines.
56   Lecture 7, p. 92.
57   Lecture 9, p. 126.
58   Lecture 10, p. 135.
59   Ibid., p. 136.
60   Ibid.
61   Ibid., p. 139.
62   Paris, *Life*, p. 188.
63   *Philosophical Magazine* 37 (1811): 392–98, 465–70.
64   [Thomas Allan], "Sketch of Mr. Davy's Lectures on Geology.
     Delivered at the Royal Institution, London, 1811. From notes
     taken by a private Gentleman," 1811.
65   Royal Institution Davy MSS. Geology Lecture 5, 1811.
66   John Davy, *Fragmentary Remains, Literary and Scientific, of
     Sir H. Davy* (London, 1858), pp. 134–35.
67   Davy, *Works*, 6: 207–16.
68   Ibid., p. 208.
69   Paris, *Life*, p. 388.
70   Davy, *Works*, 6: 213, 214.
71   Ibid., 8: 232.
72   Ibid., 5: 101.
73   Ibid., 138.
74   Ibid., 8: 186–87.
75   Ibid., 234.
76   Paris, *Life*, p. 194.
77   Ibid., p. 267.
78   John Davy, *Memoirs*, 1: 481.
79   Davy, *Works*, 7: 9.
80   John Davy, *Memoirs*, 2: 166.
81   Davy, "On the Phenomena of Volcanoes," *Works*, 6: 344–58.
82   Ibid., p. 354.
83   Ibid., pp. 357–58.
84   From a letter to his wife, dated 1 March 1829, John Davy,
     Remains, p. 311.
85   William Buckland, "An Account of an Assemblage of Fossil

Teeth and Bones," *Philosophical Transactions of the Royal
Society of London* 112 (1822): 171-236.

86 Davy, *Works*, 7: 40.
87 Ibid., 9: 297.
88 Ibid., 302.
89 Ibid., 299.
90 Ibid., 302-3.
91 Charles Lyell, *Principles of Geology*, 3 vols. (London, 1830–
    33), 1: 144. For an account of Lyell's use of Davy's posi-
    tion, see Robert Siegfried and R. H. Dott, Jr., "Humphry Davy
    as Geologist," *British Journal for the History of Science* 9
    (1976): 219-27.

LECTURE ONE

1 The unidentified second rock Davy demonstrated here was probably a piece of micaceous schist. See his related comments
   in Lecture 5, p. 65.
2 Davy's terminology was loose even for his own time. Here he
   uses "schist" to denote what was probably shale, though he
   often applies the same word to the micaceous metamorphic
   rock, as in modern usage. His frequent distinction between
   hard and soft schist presumably corresponds to our modern
   schist and shale. In the paragraph immediately above he refers to shale as soft slate, though in Lecture 8 he uses the
   term "shale."
3 This generalization apparently refers only to British mineral
   veins, since in Lecture 8 Davy specifically disclaims any
   rule for the orientation of mineral veins as applicable to
   all countries. He there notes that the greatest mineral veins
   in Britain (including his own Cornwall) trend east-west, but
   other directions are cited for South America and for Hungary.
   That the contemporary belief was that the richest tin-copper
   veins trend east-west is confirmed by William Phillips's field
   work done in 1800. "On the Veins of Cornwall," *Transactions of
   the Geological Society of London* 2 (1814): 110-60. Subsequent work has shown that silver-lead veins trend north-south.
   D. A. MacAlister, "Geological Aspects of the Lodes of
   Cornwall," *Economic Geology* 8 (1908): 363-80. The motivation
   for seeking such generalizations was practical rather than
   theoretical in Davy's time.
4 It is not clear what mineral Davy means by "spar." According
   to William Phillips, Cornish miners spoke of veins being
   "sparry" when quartz predominated, but they also called veins
   with "fluate of lime" (fluorspar, fluorite) "sparry" loads.
   "On the Veins of Cornwall," n. 3, pp. 117-18. Other minerals
   that Davy might have been referring to include feldspar,
   heavy spar (barite), or Iceland spar (calcite). Any of these
   can be present in the tin-bearing veins of Cornwall with
   which Davy was most familiar.
5 "Gossan" is a Cornish term still used for a deeply weathered,

superficial cap of a mineral vein. Gossans are generally enriched in iron oxide minerals, giving them a yellow or brown
coloration.

6 See the Preface for a discussion of Davy's use of illustrations.

7 Davy's version of the Hindu account of creation is taken almost verbatim from Sir William Jones's translations found in
"On the Gods of Greece, Italy, and India," *Asiatic Researches*
1 (1806): 221-75. Our quotation is from the 1806 edition (p.
244), though Davy presumably used the earlier 1801 edition
which was available in the Royal Institution Library. See
William Harris, *Catalogue of the Library of the Royal Institution* (London, 1809), p. 297. The following quotation is
also a very close paraphrase of Jones's version of the passage from Genesis in the same paper.

8 The best reading of the manuscript here is "Cre," though this
does not correspond to any of the variously spelled names
given for the father of Osiris. We have used "Keb" following
E. A. Wallis Budge, *Osiris* (New York, 1961). We have not
been able to locate or identify Davy's source for this version of Egyptian cosmogony. Captain Francis Wilford's paper,
to which Davy refers ("On Egypt and other Countries adjacent
to the Ca'li' River or Nile of Ethiopia: From the Ancient
Books of the Hindus," *Asiatic Researches* 3 (1806): 295-462),
does indeed argue the similarity and the presumed common origin of the Hindu, Egyptian, and Greek mythologies, but we
cannot establish that Davy's brief version was based on this
paper.

9 Davy's reference to modern scepticism toward the fall of
meteorites was most timely, for it was only in 1803 that a
report to the French Academy by Jean Baptiste Biot had fairly
well established the fall of stones from the sky as a natural
reality. A brief account of eighteenth-century disputes
about meteorites can be found in Willy Ley, *Watchers of the
Skies* (New York, 1969), pp. 232-37.

LECTURE TWO

1 Jean-Baptiste Romé de L'Isle and René-Just Haüy were the two
French mineralogists chiefly responsible for the early development of the science of crystallography.

2 But the *Encyclopedia Britannica* says of this work, "It is
certainly not authentic, and cannot be dated earlier than the
1st century B. C." *Encylopedia Britannica*, 11 ed., s.v.
"Ocellus Lucanus."

3 Davy here and below appears to accept the historical reality
of Timaeus on the authority of Diogenes Laërtius. Francis
M. Cornford, however, says that "there is no evidence for
the historic existence of Timaeus of Locri," and that the
treatise *On the Soul of the World and Nature* is a forgery
of the first century A.D., *Plato's Timaeus*, ed. Francis

M. Cornford (Indianapolis, 1959), pp. xviii-xix.

4 Here Davy presumably refers to Democritus who lived perhaps forty years after Leucippus, and today generally is regarded more highly in the development of the atomic hypothesis than is Leucippus.

5 "Fossil productions" in Davy's time still meant "anything dug up," including inorganic minerals and rocks as well as the lithified organic remains to which the term is confined largely today.

LECTURE THREE

1 Agrippa von Nettesheim, according to the *Dictionary of Scientific Biography,* was born in 1486. Both versions of the Davy manuscripts give 1463, however.

2 Theophrastus Philippus Aureolus Bombastus von Hohenheim assumed the name Paracelsus by which he is known almost universally today. His influence on chemistry and medicine was great.

3 This is the best reading we can make of an interlined passage made in Davy's hand in copy A of the lectures. The copyist's version B simply ends the sentence after the word "powers."

4 Johannes Baptista van Helmont was the last of the significant followers of the Paracelsian tradition.

5 All three of these authors are representative of late Renaissance natural philosophy and published mystical views in which alchemy played a part. Basil Valentine's fifteenth-century existence is in doubt, but Jacob Boehme and Robert Fludd were real enough, though no less mystical in their philosophy.

6 Georgius Agricola, also known as Georg Bauer, wrote the famous *De Re Metallica,* published posthumously in Basel in 1556 by Frobenius. At the time of Davy's lectures, the Royal Institution Library contained a copy of the 1561 edition by the same printer.

7 The work by Lazarus Ercker Davy refers to was catalogued in the Royal Institution Library as, Sir John Pettus's *Fleta Minor; The Laws of Art and Nature in Knowing, Assaying and Refining Metals; with a Dictionary of Metallick Words* (London, 1683). Pettus translated Ercker's work and added the dictionary.

8 For a modern account of these controversies see Frances Yates, *Giordano Bruno and the Hermetic Tradition* (New York, 1964), especially the last chapter.

9 *The Posthumous Works of Robert Hooke, M. D., S. R. S.,* ed. Richard Waller (London, 1705). Hooke, like Leonardo da Vinci and Bernard Palissey before him, and Nicholas Steno, Leibniz, and John Woodward in his own time, argued strongly against the prevailing notions of an inorganic origin for fossil shells and a single deluge origin for them. To Hooke, the fossils were remnants of earlier times from which a history

of the earth might be constructed. This idea was reconceived
and brought to detailed fruition in the decade after Davy's
lectures, chiefly by William Smith (1769-1839; DSB 12: 486),
Georges Cuvier (1769-1832; DSB 3: 521), and Alexandre
Brogniart (1770-1847; DSB 2: 493).

10   Johann Joachim Becher. Davy is mistaken here, for Becher
wrote voluminously on a wide variety of topics, especially
economics and public administration. He even attempted to
formulate a universal language. His chief focus was on chemistry, which was then heavily influenced by alchemical and
Paracelsian traditions. The work Davy refers to carried the
Latin title *Actorum laboratorii chymici Monacensis, seu
Physicae subterraneae libri duo* (Frankfurt, 1669) but is
more generally and conveniently known as *Physica subterranea*.

11   Georg Ernst Stahl is famous in medicine for his vitalistic
views and in chemistry for his creation of the phlogiston
theory. Davy discusses Stahl briefly in Lecture 8.

12   According to Francis C. Haber, only a two page resumé of
Wilhelm Gottfried Leibniz's *Protogaea* appeared in 1693 in the
learned review published in Leipzig, *Acta Eruditorum*. The
full work did not appear until a Latin edition was published
posthumously in 1749. *The Age of the World: Moses to Darwin*
(Baltimore, 1959), p. 85.

13   *G. W. Leibnitii opera omnia,* ed. Louis Dutens (Geneva, 1768).

14   William Whiston published his biblically based cosmology in
New Theory of the Earth (London, 1696).

15   Thomas Burnet published the original Latin version in 1681
and the English edition in 1684. *Sacred Theory* was one of
the earliest and most vigorous attempts to reconcile Genesis
with geology. Similar attempts by later writers did not entirely displace it, for it was reprinted throughout the
eighteenth and in the early nineteenth centuries. See Haber,
Age of the World, pp. 71-84.

16   John Keil was one of the more vigorous promoters of Newtonian
mechanical philosophy. His criticism of Burnet's *Sacred
Theory* appeared in *An Examination of Dr. Burnet's Theory of
the Earth: Together With Some Remarks on Mr. Whiston's Theory
of the Earth* (Oxford, 1698). Keil, like Burnet and Whiston,
sought a reconciliation of Scripture with the new science,
but he thought their natural theology dominated Scripture too
much and sought to make natural theology subordinate to Scripture. DSB 7:275.

17   Georges-Louis Leclerc, Comte de Buffon. The fullest account
of his cosmogony and history of the earth is found in his
Époques de la nature (Paris, 1779).

18   Though Davy refers to the work of H. B. de Saussure in other
contexts (e.g., see Lecture 5, p. 65) he was apparently una-
ware of that investigator's attempts in the 1780s to reproduce granites and basalts from their molten states. Like the
later efforts of Sir James Hall and Gregory Watt, de
Saussure's efforts failed to reproduce the crystal structure
of the original rocks. Cf. Archibald Giekie, *The Founders*

of Geology (New York, 1902; reprint ed., New York, 1962),
p. 190.

19 This is the only place where Davy mentions an actual value of
time related to geological process. But 4,000 years apparently refers to the length of historical time within which
diminution could be observed. It seems unlikely that he intended this value to represent the age of the earth itself,
for even the traditional biblical age allowed 6,000 years.

LECTURE FOUR

1 Benoît de Maillet founded his system of the steady diminution
of the sea on the studies and observations he made while on
diplomatic duty in Egypt. His reliance on Cartesian natural
law appeared dangerously materialistic and the work was published first in 1748, ten years after the author's death.

2 Edmund Halley, though justly most famous for his contributions to mathematical and observational astronomy, was also
active in many other areas of science. Davy is referring
here to three brief papers Halley published in the *Philosophical Transactions* in the early 1690s, in which he established that the rate of evaporation from the surface of the
Mediterranean was more than sufficient to account for the
return of water to the sea by its major rivers. Davy's
language implies a rigor greater than Halley would have
claimed.

3 John Woodward was an avid collector of fossils, which he un-
equivocally identified as organic remains. But characteristically for his time, he attributed their deposition to
the universal Deluge. His *Essay Towards a Natural History
of the Earth* (London, 1695) was widely read.

4 John Whitehurst, author of *Inquiry into the Original State
and Formation of the Earth* (London, 1778), is the only one
of the authors Davy mentions in this paragraph who was pri-
marily a geologist. All the others, Joseph Black, Henry
Cavendish, Joseph Priestley, and Torbern Bergman were pri-
marily chemists.

5 "Toadstone" is an old name used chiefly in Derbyshire for
amygdaloidal basalt, a variety with many original cavities
which are later filled with some secondary mineral. Such
basalts sometimes weather to a knobby surface, perhaps in-
spiring the name.

6 Jean André Deluc was a native of Geneva, but after 1773 he
spent the greater part of his time in London, where he held a
post as reader to Queen Charlotte. His *Letters to the Queen*,
referred to by Davy below, was originally published as
*Lettres physiques et morales sur les montagnes et sur
l'histoire de la terre et de l'homme: Adressées à la reine de
la Grande Bretagne* (The Hague, 1779). He wrote voluminously
on many topics in chemistry and physics, but his main interests were in natural history, and he was strongly moti-

vated by a desire to reconcile Genesis with geology.

7   Johan Gottschalk Wallerius, Professor of Chemistry at the
University of Upsala after 1750, was a prolific writer on a
wide variety of scientific topics, chiefly chemistry and
mineralogy. His cosmology was presented in the Latin work
Meditationes Physico-chemicae de origine mundi (Stockholm,
1779), which was apparently not available to Davy who got
his account of it from Howard. See text below.

8   Philip Howard, [No dates for Howard are given in Rev. John
Kirk's *Biographies of English Catholics in the Eighteenth
Century* (London, 1909), p. 131.] The correct title of the
book Davy refers to here is *Scriptural History of the Earth
and of Mankind* (London, 1797), which is chiefly a detailed
refutation of the schemes of Buffon and of Hutton. See
Francis C. Haber, *The Age of the World: Moses to Darwin*
(Baltimore, 1959), pp. 192-94.

9   James Hutton, the central figure in whose name the Plutonian
view of the origin of rocks was defended, published a full
account of this view in *Theory of the Earth* (Edinburgh,
1795). As is clear from Davy's account below, "Plutonism" is
something of a misnomer, for Hutton's system rests fundamen-
tally on an equilibrium between the degrading forces of water
and the creating forces of heat. The significance of heat in
Hutton's theory was probably sufficient to invite the term
"Plutonism" as a contrast to Werner's almost total Neptunism.

10 Antonio-Lazzaro Moro was an early Plutonist who was severely
critical of the elaborate conjectural schemes of Burnet and
Woodward.

11 Abraham Gottlob Werner was the chief author of what became
known by the end of the eighteenth century as the Neptunian
view of the origin of the rocks. It was largely through the
efforts of his students at the Freiburg School of Mines that
his ideas on the formation and succession of rock strata were
widely and vigorously advocated and defended. Werner coined
the term "geognosy" to emphasize his own insistence that the
science dealt only with earth's upper layers, the only part
which could be directly observed. Richard Kirwan, an Irish
mineralogist and chemist, generally was considered a Neptunist, for though he had no direct connection with Werner and
held views in contradiction to those of Werner, his desire
to accommodate the earth's formation and structure to the
biblical accounts of creation and the Deluge made him violently anti-Huttonian and thus, roughly, among the Neptunians.
His major anti-Huttonian work was his *Geological Essays*
(London, 1797).

12 Hall's experiments, briefly described below by Davy, were
undertaken to meet certain objections to the Huttonian theory
of the earth, to which Hall had been at least partially converted by Hutton himself.

13 John Playfair was Professor of Mathematics and later Professor of Natural Philosophy at the University of Edinburgh.
His *Illustrations of the Huttonian Theory of the Earth*

(Edinburgh, 1802) was the chief means by which Hutton's theory became widely known. Davy met Playfair during his visit to Edinburgh in the summer of 1804.

LECTURE FIVE

1 "New Holland" was the name still used in Davy's time for what is now called Australia, for it was the Dutch who first explored the continent in the seventeenth century and claimed it as their own. British political dominance came late in the eighteenth century, the first colony being established at Botany Bay in 1798.
2 The term "schorl" is now confined to black tourmaline, but in the eighteenth century, it was applied to many dark prismatic mineral species. Davy, here and elsewhere in these lectures, appears to be using "schorl" to identify loosely another dark mineral, probably hornblende. Microscopic identification techniques were not yet available, and chemical analysis was neither reliable nor standardized, so distinctions were not easy to make nor was the need for distinction recognized.
3 Mull is one of the islands of the Inner Hebrides Davy visited in the summer of 1804.
4 Logan Rock is on a coastal headland about 6 kilometers southeast of Land's End in Cornwall.
5 Quartz rock probably refers to quartzite, either a very firmly indurated quartz sandstone or a metamorphosed sandstone.
6 Davy's discussion of limestone and marble illustrates the absence of any general concept of metamorphosis, which became widely used only after 1830. The eighteenth-century concept of primitive rock referred to those formed at creation, while secondary rocks were those formed during the earth's subsequent history, chiefly from the detritus of the primitive rocks. Thus the presence of fossils necessitated classifying rocks as secondary and generated an arbitrary distinction between them and rocks with identical composition and crystal form, but without fossils. Hence primitive and secondary limestone, primitive and secondary marble, etc.
7 Horace Bénédict de Saussure made many detailed studies of the Alps between 1774 and 1787 from his home in Geneva. His careful observations were published in four volumes under the title of *Voyages dans les Alpes* (Neuchatel, 1779-96).
8 St. Michael's Mount is a rocky island on the northeast side of Mount's Bay only 2.5 kilometers east of Davy's native Penzance. The Mount and its bay were of great personal significance to Davy, figuring in much of his youthful poetry. Here, however, the reference illustrates his laudable practice of using examples with which he is personally familiar.
9 Davy's argument against the Huttonian theory here is not entirely clear, but it seems to go as follows: If the primitive rocks have been formed by fire from previously existing rocks

containing elemental carbon, then surely the carbon, being non-volatile, would still be found within the body of the new rocks after they had cooled and solidified. Since we do not find carbon in primitive rocks, something is wrong with Hutton's theory.

10 Johan Gottlob Lehmann was a prolific writer on mineralogical subjects and the structure of mountains. He was influential in the founding of the Bergakademie in Freiberg in 1765. The book Davy cites is *Abhandlung von den Metall-Muttern und der Erzeugung der Metalle* (Berlin, 1753). Davy did not read German nor did the Royal Institution have a copy of this work.

11 Davy here is echoing a conservative eighteenth-century chemical rule, that the composition of a substance is established only after it has been determined by both analysis and synthesis. The ability to synthesize granite, gneiss, schist, and other igneous and metamorphic rocks was not achieved until the twentieth century. Not only high temperature, but also moderately high pressure and means for controlling the water content, are required.

LECTURE SIX

1 Robert Plot, *The Natural History of Oxfordshire* (Oxford, 1677). The book, a kind of traveler's guide to natural and cultural curiosities, was very popular and had many imitators.

2 Sir Thomas Browne was a physician and an energetic amateur scientist and naturalist. His *Pseudodoxia epidemica* (London, 1646) contains most of his writings on scientific topics.

3 Edward Lhwyd was a distinguished naturalist of his time and Keeper of the Ashmolean Museum in Oxford.

4 Robert Hooke's "Lectures and Discourses of Earthquakes" constituted the largest portion of his posthumous works. See *The Posthumous Works of Robert Hooke, M. D., S. R. S.*, ed. Richard Waller (London, 1705).

5 John Ray was the outstanding British naturalist of his time, and his *Wisdom of God Manifested in the Works of Creation* (London, 1691) became the pattern for British natural theology for the following century.

6 Déodat de Gratet de Dolomieu was an accurate observer of geological formations who traveled widely in the mountains of southern Europe and among the islands of the western Mediterranean, publishing his reliable descriptions in extended volumes. The mineral, dolomite, and the Dolomites range of the Alps are named for him.

7 Johann Heinrich Pott, Torbern Bergman, Carl W. Scheele, and Martin Heinrich Klaproth all made significant contributions to the techniques for the chemical analysis of minerals. Pott and Klaproth were Germans; Bergman and Scheele were Swedes.

8 The painting Davy refers to is based on an engraving from

H. B. de Saussure, *Voyages dans les Alpes,* 4 vols. (Neuchâtel, 1779-96), 1: 3.

9   Charles Hatchett was a skilled chemical analyst and one of the managers of the Royal Institution.

10 At this point Davy apparently held up or pointed to the book by Johann Scheuchzer, *Homo diluvii testis* (Tiguri, 1726), which contained a drawing of the skeleton of the man who was supposed to have witnessed the Deluge. Scheuchzer was a Swiss physician and naturalist who, like a few others of that time, defended the view that fossils were indeed relics of formerly living creatures whose remains were deposited by the Deluge. In 1805 there was no work by Scheuchzer listed in the *Catalogue of the Library of the Royal Institution.* See Melvin E. Jahn, "Some Notes on Dr. Scheuchzer and on *Homo diluvii testis,*" *Toward a History of Geology,* ed. Cecil J. Schneer (Cambridge, Mass., 1969), pp. 192-213.

11 Although Davy gives no indication of the source of this report, the remains of organisms contained in alleged basalt were being cited by others about this time as evidence for its aqueous rather than igneous origin. In particular, Playfair discussed in his *Illustrations of the Huttonian Theory of the Earth* (Edinburgh, 1802) specimens from Portrush, a few miles west of the Giant's Causeway on the north coast of Ireland. He described them as shells and impressions (mostly of *cornu ammonis*) occurring in a dark, fine hornstone or chertlike rock that had been indurated by, and confused with, adjacent fine-grained basalt (par. 252, pp. 286-88). Davy had read Playfair's *Illustrations* and must have been aware of the possibility of this kind of confusion. Davy visited Portrush in 1805 and examined the basalt there, yet in his geology lectures of 1811, he stated that basalt "sometimes contains shells, as in the fine-grained variety which occurs at Portrush." From Thomas Allan's 1811 account, "Sketch of Mr. Davy's Lectures on Geology.   Delivered at the Royal Institution, London, 1811. From notes taken by a private Gentleman," p. 25.

12 Sir Joseph Banks was a naturalist made famous by his voyage around the world with Captain James Cook on the *Endeavour* in 1768-71. He was president of the Royal Society from 1780 until his death in 1820 and played a significant supporting role in the founding and early governing of the Royal Institution. Bank's account of Fingal's Cave appeared in Thomas Pennant's *Tour in Scotland, and Voyage to the Hebrides,* 3 vols. (London, 1774). Uno von Troil, a Swedish clergyman, accompanied Banks on this tour, which also went on to Iceland. He published his own account of Staffa and Fingal's Cave in his *Letters on Iceland. . .* (London, 1780), in which he included Banks's account following his own. For a brief biographical account of von Troil in Swedish see volume 8 of *Svenska Män och Kvinnor, Biografisk Uppslagsbok* (Stockholm, 1955). Davy refers here to the Hon. Mrs. Murray (afterward Aust), *A Companion and Guide to the Beauties of Scotland*

and the Lakes of Westmoreland, &c., 2 vols. (London, 1799-1805).

13 Thomas Webster had at one time been Clerk of the Works at the Royal Institution but left the Institution's employ in 1802. He was a draftsman and artist of some skill, having been largely responsible for the design of the lecture room of the Royal Institution, justly known even today for its excellent acoustical and visual qualities. Webster later was employed professionally by the Geological Society of London and accomplished some pioneering work in geological mapping before becoming a lecturer on geology after 1827.

LECTURE SEVEN

1 Hall's first published account of these experiments appeared only after Davy had visited him in Edinburgh. Sir James Hall, "Account of a Series of Experiments, Showing the Effects of Compression in Modifying the Action of Heat," *Journal of Natural Philosophy, Chemistry, and the Arts* 13 (1806): 328-43, 381-405; 14 (1806): 13-22, 113-28, 196-212, 302-18.

2 Joseph Black, "An Analysis of the Waters of Some Hot Springs in Iceland," *Transactions of the Royal Society of Edinburgh* 3 (1794): 95-126.

3 Until shortly after his mother's death in 1800, James Louis Macie Smithson went by her name of Macie. He was a competent analytical chemist but is best remembered for leaving his fortune to the United States for the establishment of the Smithsonian Institution. Davy refers to the paper, "Account of Some Chemical Experiments on *tabasheer,*" *Philosophical Transactions of the Royal Society of London* 81 (1791): 368-88.

4 The paper Davy refers to is Hatchett's "Observations on the Change of some of the Proximate Principles of Vegetables into Bitumen; with Analytical Experiments on a peculiar Substance which is found with the Bovey Coal," *Philosophical Transactions of the Royal Society of London* 94 (1804): 385-410.

5 George B. Greenough [1778-1855; DSB 5:518], when visiting the Salisbury Crags the summer following Davy's lectures, raised the same doubts as Davy expressed here that the hardness of the sandstone was owing to heated contacts with the once molten basalt. See excerpts from Greenough's Journal printed in Martin J. S. Rudwick's "Hutton and Werner Compared: George Greenough's Geological Tour of Scotland in 1805," *British Journal for the History of Science* 1 (1962): 117-35.

6 Gregory Watt (1772-1804), second son of James Watt, carried out his large scale experiments in his father's Soho Engineering Works in Birmingham. He and Davy had been close personal friends. "Observations on Basalt, and on the Transition from the vitreous to the stony texture, which occurs in

the gradual refrigeration of melted basalt; with some geological remarks," *Philosophical Transactions of the Royal Society of London* 94 (1804): 277-314. No separate biographical source is available.

7   Jean André Deluc, *Lettres sur l'Histoire Physique de la Terre* (Paris, 1798). These letters were originally published in the *British Critic* from 1793 to 1795.

8   Possibly this is Sir George Paul (1746-1820). The philanthropic son of a successful woolen cloth manufacturer, Paul devoted much of his life to the improvement of prisons and to aid for the poor, chiefly in the neighborhood of Gloucester. *Dictionary of National Biography*, s.v. "Paul, George."

LECTURE EIGHT

1   Alvaro Alonso Barba, *El arte de los metales* . . . (Madrid, 1640). The work was first translated into English by Edward Montagu, earl of Sandwich, with the title, *The First Book of the Arts of Mettals* (London, 1670). The Royal Institution Library contained the English edition of 1740.

2   William Pryce, M. D., *Mineralogia Cornubiensis* (London, 1778). See *Dictionary of National Biography*, s. v. "Pryce."

3   William Cookworthy (1705-80), a successful druggist in Plymouth, was the discoverer of the Cornish china clay, and he unsuccessfully attempted the manufacture of porcelain. He is mentioned in Pryce's *Mineralogia Cornubiensis*, where Davy probably learned of him, as a believer in the efficacy of the divining rod. See the *Dictionary of National Biography*, s. v. "Cookworthy."

4   Georg Ernst Stahl's *Specimen Beccherianum* (Leipzig, 1703) was appended to Stahl's edition of Becher's *Physica Subterranea*. At the time of Davy's lectures, the Royal Institution Library had the 1738 edition of this work.

5   Johann Friedrich Henckel had been a medical and a chemical student of Stahl's when he founded a chemical laboratory of his own in Freiberg. His success in mineral analysis and the enthusiasm of his followers led to the founding of the Bergakademie in Freiberg in 1765. There was no book by Henckel in the Royal Institution Library in 1805.

6   Abraham Gottlob Werner, *Nouvelle Theorie des Formation des Filons: Application de cette Theorie à l'Exploitation des Mines* (Paris, 1802), a translation of a work originally published in German in 1791. Only the French edition was in the Royal Institution Library in 1805.

LECTURE NINE

1   The word is missing from the copyist's version, and the page is torn out of the copy written in Davy's hand. We have supplied "Phlegraean Fields" because this name is often

associated in antiquity with the battle of the gods, though
its location is uncertain. In modern times the name has become associated with a volcanic area near Naples. Cf. note 3
below.

2 *The Odes of Pindar*, trans. Gilbert West (London, 1751). The
passage given occurs in the fifth decade of the First Pythean
Ode, as found in Robert Anderson, ed. *British Poets*, 13 vols.
(London, 1794-95), 12: 311. The page is missing from the
copy of the lecture written in Davy's hand, and the copyist's
version of these lines differs slightly from the published
version which we have reproduced here.

3 Sir William Hamilton was a British diplomat, amateur scientist, and husband of Emma Hamilton, Lord Nelson's longtime
mistress. Hamilton published a series of observations on volcanoes in the *Philosophical Transactions of the Royal Society*
between 1767 and 1795. The book referred to is *Campi
phlegraei: Observations on the Volcanoes of the Two Sicilies*
(Naples, 1776). There was a copy in the Royal Institution
Library at the time of Davy's lectures.

4 Lazzaro Spallanzani, an Italian, worked tirelessly on experiments, chiefly in the life sciences, and was a perennial
traveler. An English translation of his account of his
1788 journeys was available in the Royal Institution Library
in 1805. *Travels in the Two Sicilies and Some Parts of the
Apennines*, 4 vols. (London, 1798).

5 Scipione Breislak was an Italian of German extraction and a
priest. The Royal Institution Library in 1805 possessed only
one of Breislak's several books, *Voyages physiques et
lythologiques dans la Campanie*, 2 vols. (Paris, 1801).

6 Johann Jacob Ferber (1743-90) was a Swedish mineralogist and
chemist whose *Travels through Italy in the Years 1771 and
1772* [trans. G. E. Raspe (London, 1776)] was available to
Davy in the Royal Institution Library, indexed in the Catalogue under the translator's name. Ferber is not listed in
the DSB, but see J. C. Poggendorf, *Biographish-literarisches
Handwörterbuch zur Geschichte der exacten Wissenschaften*
(Leipzig, 1873), s. v. "Ferber."

7 The work by Dolomieu that Davy refers to was probably *Mémoire
sur les îles Ponces, et catalogue raisonné des produit de
l'Etna* (Paris, 1788), but perhaps he also meant *Voyages aux
îles de Lipari fait en 1781* (Paris, 1783). Both works were
in the Royal Institution Library in 1805.

8 Alexander von Humboldt was a naturalist and world traveler
whose interests were genuinely as broad as those implied in
the title of his major popular work, *Kosmos: Entwurfeiner
physischen welt beschreibung*, 5 vols. (Stuttgart-Tübingen,
1845-62). The Royal Institution Library contained no books
by Humboldt. Davy probably read Humboldt's brief paper,
"Curious Particulars respecting the Mountains and Volcanoes,
to the Effect of the late Earthquakes in South America,"
Journal of Natural Philosophy, Chemistry, and the Arts, 6
(1803): 242-47; abridged from the *Magazin Encyclopédique*.

9   Giulio Cesare Braccini, *Dell' incendi fattosi nel Vesuuio a xvi Dicembre M.DC.XXXI* (Naples, 1632). There was no copy of this work in the Royal Institution Library. Braccini is not otherwise known as a scientist, but his fortuitous presence in Naples in 1631 allowed him to write a detailed and literate account of his careful observations. Cf. *Dizionario Biografico degli Italiani,* 20 vols. to date (Rome, 1972-), 13: 631-32, s. v. "Braccini." No dates are given for his birth or death.

10 That the friend was George Bellas Greenough has been confirmed through the kindness of Mr. P. J. Gautrey of the Cambridge University Library who compared the passages given by Davy with Greenough's uncatalogued notebooks now in Cambridge University's manuscript collections.

11 Davy's translation and paraphrasing is from Breislak, *Voyages physiques et lythologiques,* 1: 184-85.

12 Francesco Ferrara was Professor of Natural History at the University of Palermo and later Professor of Mathematics and Physics at the University of Catania. Davy made no reference to Ferrara in these lectures other than this indirect one from Greenough's account. Ferrara is not listed in the DSB; see Poggendorf, *Biographisch-literarisches,* s. v. "Ferrara."

13 The following passage is treated editorially as a direct quotation even though there are many omissions from the original along with appropriate paraphrases and connecting passages by Davy, none of which he indicated in any way. The content is quite faithful to the original as found in Spallanzani's *Travels in the Two Sicilies,* 1: 246-49.

14 The following passage is treated editorially as explained in note 13. It is derived from Sir William Hamilton, "Account of the Late Eruption of Mount Vesuvius," *Philosophical Transactions of the Royal Society of London* 85 (1795): 73-116.

15 Heneage Finch, second earl of Winchelsea. *A True and Exact Relation of the Late Prodigious Earthquake and Eruption of Mount Etna, or, Monte-Gibello; as it came in a letter written to His Majesty from Naples by the Right Honourable the Earle of Winchelsea,* etc. (London, 1969). We have been unable to locate a copy of this work to check the accuracy of Davy's quotation. Presumably he handled it as he did the earlier quotations in this lecture.

16 Nicolas Lemery was the author of *Cours de Chimie,* which became the most popular chemistry text of its time. It went through more than ten French editions from 1675 to 1747 and was translated into English, German, Dutch, and Spanish. According to James Partington, the experiment Davy refers to was first reported by Lemery in 1690 in the fifth edition of his famous text. *A History of Chemistry,* 4 vols. (London, 1961-70), 3: 36.

17 Dr. Robert Plot, *The Natural History of Staffordshire* (Oxford, 1686). Plot's version of the story can be found on page 142.

LECTURE TEN

1 John Stephens, "An Account of an Uncommon Phenomenon in
   Dorsetshire," *Philosophical Transactions of the Royal Society
   of London* 52 (1761): 119-23.
2 John Williams, *The Natural History of the Mineral Kingdon*, 2
   vols. (Edinburgh, 1789).
3 Davy made a mistake here, for the author of *Voyage to the
   Lipari Islands* was Dolomieu, not Spallanzani, and the book
   contains no plates in any case. But Spallanzani's *Voyages
   dans le deux Siciles* does contain a plate of the crater of
   Stromboli and undoubtedly this is the one Davy refers to.
   See Figure 10.1.

Index

Numbers in italic type indicate pages where illustrations appear.

DESIGNED BY IRA NEWMAN

MANUFACTURED BY THOMSON-SHORE, INC.

DEXTER, MICHIGAN

Library of Congress Cataloging in Publication Data
Davy, Humphry, Sir, bart., 1778-1829.
Humphry Davy on geology.
Includes bibliographical references and index.
1. Geology--Addresses, essays, lectures.
I. Siegfried, Robert, 1921-
II. Dott, Robert H., 1929-
III. Title
QE35.D26 1980   550   79-5022
ISBN 0-299-08030-7